Detection and measurement of hazardous gases

Detection and measurement of hazardous gases

Edited by

C. F. Cullis and J. G. Firth

HEINEMANN
London · Exeter, NH

Published in Great Britain by
Heinemann Educational Books Ltd
22 Bedford Square, London WC1B 3HH

LONDON EDINBURGH MELBOURNE AUCKLAND
HONG KONG SINGAPORE KUALA LUMPUR NEW DELHI
IBADAN NAIROBI JOHANNESBURG
EXETER (NH) KINGSTON PORT OF SPAIN

Published in the USA by
Heinemann Educational Books Inc.
4 Front Street, Exeter, New Hampshire 03833

ISBN 0 435 71030 3

British C.I.P. Data

Cullis, C. F.
Detection and measurement of hazardous gases.
1. Gases, Asphyxiating and poisonous —
Safety measures
I. Title II. Firth, R.
363.7'387 TD885

ISBN 0-435-71030-3

Library of Congress C.I.P. Data
Main entry under title:

Detection and measurement of hazardous gases.
Includes bibliographical references.
Contents: History and law/J. H. Burgoyne—Measurement of flammable gases and vapours/J. G. Firth—Oxygen deficiency/L. R. Cooper—etc.
1. Gases, asphyxiating and poisonous—Measurement.
I. Cullis, C. F. II. Firth, J. G.
TD890.D48 604.7 81-2785
ISBN 0-435-71030-3 AACR2

First published 1981

Designed and produced for Heinemann Educational Books by
Science Reviews Ltd, 3/4 St Andrew's Hill, London EC4V 5BY
Typeset by Advanced Scientific Press Ltd, Cambridge, England
Printed by Richard Clay (The Chaucer Press) Ltd, Bungay, Suffolk

Preface

The number and amounts of industrial materials which can be present as gases or vapours in work-place atmospheres is increasing rapidly. The concentration levels of many of these substances, which have to be measured in the working environment, are being decreased as new legislation becomes effective. The complexity of work processes is, however, constantly increasing, resulting in many cases in an increased possibility of escape of substances into the atmosphere. This has necessitated in recent years the rapid expansion of the numbers and types of instruments, devices and procedures for measuring the amounts of potentially toxic and flammable gases and vapours in work-place atmospheres.

Many of these instruments, devices and procedures involve common techniques even though they are used for widely different applications. These techniques have different degrees of applicability to particular circumstances. The object of this book is to help people who are concerned with measuring hazardous gases to understand more fully the advantages and limitations of particular techniques when applied to individual situations. The book is not intended to be an encyclopaedia of all possible measuring methods and means of interpreting such measurements but discusses the techniques which are most widely used at present and are most likely to be used in the future. It is written by people who are involved in both the development and application of techniques for measuring hazardous gases and it is hoped therefore that it will help the reader to decide the appropriate technique to use for a particular problem and, of equal importance, to understand what the measurements mean in terms of the hazard present.

C. F. Cullis
J. G Firth

Contributors

J. H. BURGOYNE (History and law) is consultant to the firm Dr J. H. Burgoyne and Partners, consulting scientists and engineers, which he founded and from which he retired in 1978. He has published upwards of 100 papers and articles on fire and explosion hazard problems. He has served on several government committees and was chairman of the Committee of Inquiry into Offshore Safety. He was awarded the CBE in 1980.

J. G. FIRTH (Measurement of flammable gases and vapours) is deputy director of the Occupational Medicine and Hygiene Laboratories of the Health and Safety Executive. His responsibilities include the development of new sampling and analytical techniques for measuring toxic gases and vapours and the development of sensors and field instrumentation for the measurement of both flammable and toxic gases and vapours. He is the author of numerous papers and patents in this field and is a member of national and international committees concerned with techniques and standards for hazardous gas measurement.

L. R. COOPER (Oxygen deficiency) is head of Instrument Research Branch of the National Coal Board's Mining Research and Development Establishment. His responsibilities include the development of instruments and sensing techniques for use in coal mines and he has a special interest in instruments for gas detection and the environmental properties of coal mines.

B. MILLER and P. O. KANE (Monitoring toxic gases in the workplace) are both with the Mond Division of Imperial Chemical Industries Ltd. Miller joined ICI in 1954 and has worked in the Analytical Chemistry Group of the Division's Research Department. For the past 15 years he has specialised in method development and application for solving industrial hygiene and environmental problems for the Division and its customers. He is a member of the British Occupational Hygiene Society. Kane is a research scientist whose duties include electrochemical analytical method development. As a former member of the Chemical Industries Association VCM Analysis and Monitoring Committee, he contributed to the Association's book: *VCM – A Plant Manual.*

D. T. COKER (Personal monitoring) is an industrial hygienist for Esso Europe covering production, transportation and marketing of petroleum and petrochemicals. He has been deeply involved in vapour monitoring method developments and has published and lectured widely on the subject. He is a member of the British Occupational Hygiene Society Technology Committee, Chemical Industries Association Hygiene Committee, the Health and Safety Executive Vapour Monitoring Committee and various working groups involved in standardising monitoring methods.

R. P. HARVEY (Statistical aspects and air sampling strategies) is a senior engineering inspector of factories with the Occupational Hygiene Consultant Section of the Health and Safety Executive where he specialises in the engineering control of toxic gases, fumes and dusts and in the application of statistical methods to problems of measurement in the field of occupational hygiene.

A. F. SMITH (Standard atmospheres) is head of Training, Information and Publicity at th[illegible] Laboratory of the Government Chemist where he was formerly in charge of a s[illegible]ivision concerned with the determination of toxic substances in air. He has c[illegible]buted a number of papers dealing with the development of field tests for toxic s[illegible]nces in air to the *Analyst* and has also represented the Laboratory on a number o[illegible]mittees dealing with his field of work.

C. [illegible]LLIS (co-editor with J. G. Firth of this volume) is professor of Physical Ch[illegible]try and head of the Chemistry Department of the City University, London; he [illegible]o pro-vice-chancellor of the university. His principal research interests are the [illegible]nistry of combustion and oxidation reactions and he is the author of some 200 [illegible]ginal scientific papers in this field.

The opinions expressed by the authors are their own and are not necessarily those of their organisations.

Contents

Introduction

C. F. Cullis and J. G. Firth

Industrial development throughout the world in the 1960s and 1970s has taken place against a background of increasing concern for its possible deleterious effects on the environment and for the health and safety of the work force. This concern is not new and numerous examples exist throughout history where men of learning have observed and considered the effects of working practices on the health and safety of the worker. Hippocrates in the fourth century B.C. recognised the existence of lead toxicity in the mining industry, Ulrich Ellenbog in 1473 published a pamphlet on occupational diseases and Bernadino Ramazzini in 1700 published the first full treatise on occupational diseases. Governmental concern for the welfare of work people began with the rapid increase in the incidence of industrial disease and in the number of accidents occurring in Great Britain during the eighteenth and nineteenth centuries. This gave rise to the Factories Act of 1833 and the setting up of various inspectorates to enforce the provisions of this and subsequent legislation. The increasing social awareness in the more industrially advanced countries of the world during the last twenty years has been reflected by the passing by different governments of various Acts, which have attempted both to increase the health and safety standards in industry and to rationalise legislation which already existed for this purpose. This has given rise in the United States to the Metal and Non-Metallic Mines Safety Act of 1966, the Federal Coal Mine Health and Safety Act of 1969, the Occupational Safety and Health Act of 1970 and the Federal Mine Safety and Health Act of 1977. In the United Kingdom, the Health and Safety at Work Act of 1974 rationalised existing legislation relating to health and safety at work, defined the responsibility of the employer and the employee and brought together the different inspectorial bodies, which were then in being, into the Health and Safety Executive.

The objective of all this legislation and its enforcement is primarily to prevent industrial accidents and disease. An important instrument in achieving this is the setting up of various standards which are regarded as providing an acceptable level of safety and the monitoring of the workplace to ensure that these standards are achieved and maintained. Modern industry is using increasingly large amounts of an ever-widening range of substances which, at normal temperatures, have vapour pressures such that they form mixtures with air that will transmit a flame. Similarly, industry uses increasingly large numbers and quantities of substances that are toxic to human beings. It is estimated that in the United States there are in regular use in industry more than 28 000 different toxic substances, a large number of which have vapour pressures above the levels which can be safely breathed by man. Hence, the last twenty years have witnessed the increasing employment of systems and methods of measuring the concentration levels of flammable and toxic vapours in the working environment, in order to guard against and indicate their accidental release and to ensure that, under normal working conditions, the amounts in the atmosphere are kept within safe levels. The necessity for some working activities to be carried out in confined spaces, parts of which are only poorly ventilated, for example coal mines and sewers, and the need to enter enclosed spaces, such as storage tanks, for maintenance purposes, produces a requirement to ensure that the atmosphere in these areas contains sufficient oxygen to sustain life. Thus, the requirements for gas and vapour monitoring fall into three broad areas: flammable vapour monitoring, toxic vapour monitoring, and oxygen deficiency monitoring.

Almost all harmful gases and vapours have concentration levels which can, very roughly, be regarded as safety limits. Thus, all flammable gases have a concentration level in air called the Lower Explosive Limit (LEL) below which they will not transmit a flame. This concentration level can be measured for individual gases and vapours and for mixtures of them. Although these concentrations vary somewhat with the method of measurement, for example downwards or upwards propagation of the flame, they are well-defined physical properties of the vapour and generally lie between 0.5 and 15% v/v.

The safe concentration levels for toxic vapours are, however, much less precise. The response to a particular toxic vapour varies widely for different people. In particular, the long-term (chronic) harmful effects of a particular vapour are difficult to measure and must of necessity generally be extrapolated from animal experiments. The American Conference of Government Industrial Hygienists has for many years annually published Threshold Limit Values for about 500 toxic gases and most people can be exposed to these for eight hours each day for five days each week for a total of 40 years, i.e. an average working lifetime, without suffering noticeable harmful effects. These values are set in the light of the best available opinion and evidence at the time and are necessarily subject to change as more evidence for a particular vapour becomes available. In particular, a steadily increasing number of vapours, previously thought to be safe, are being shown to have long-term harmful effects. This appears to be particularly true for organic chlorine compounds. In contrast with long-term harmful effects, the short-term (acute) toxic effects of individual vapours are more precisely defined. The concentration levels which produce short-term effects are generally much higher than those involved in long-term effects and for many gases and vapours, unless the exposure has proved fatal, the effects are reversible. This is also the case for the effects of oxygen deficiency. Normal air at atmospheric pressure contains about 20.8% v/v of oxygen and, for the average person, concentration levels need to drop below 16% v/v before physiological effects begin to be observed.

The presence of hazardous gases in the atmosphere requires two types of remedial action. Dangerous concentrations of flammable gases, sharp decreases in oxygen concentration, and levels of toxic gases which can have an acute effect all require immediate action. The presence of toxic gases in concentrations at which only chronic effects are likely to be observed do not require the same urgent rectification. For these more acute risks, instrumentation has been developed which monitors continuously and can be sited at strategic locations in a work area; for example, places where leakages are likely to occur or, in the case of flammable gases, sites of well-defined igniting sources. Toxic gas or oxygen monitors are best carried within the breathing zone of the person at risk. Gases at concentration levels that may cause chronic effects are still dealt with mainly by techniques involving the sampling of the gas into a suitable sorbent and its subsequent analysis by laboratory-based techniques.

Although, in principle, many techniques can be employed to measure flammable gas concentrations, those involving calorimetry are most widely used on account of the fortunate existence of a relationship between the LEL of a wide range of such gases and the heat of oxidation. The fact that the calorific value of a gas–air mixture at the LEL is almost constant for a very wide range of gases has resulted in the development of small catalytic sensors which can be incorporated in simple monitoring systems and respond to changes in flammable gas concentrations within times of the order of a few seconds. These sensors can be used in portable instruments for inspection purposes as well as in fixed monitoring systems containing hundreds of these sensors distributed at strategic points throughout the workplace. The most serious defect of these sensors is the 'poisoning' of the catalyst produced by some types of vapour such as silicones, which results in loss of sensitivity of the sensors and a 'failure to danger' situation. The probable development of sensors which are resistant to such poisons will widen the use of flammable gas detection instruments and may well lead to the use of detection as a primary defence against gas explosions. At the present time monitoring is used to supplement protective systems, such as flame-proof enclosures and flame traps. The rapid and reliable detection of flammable gases at concentrations well below the LEL, combined with the rapid switching off of electrical power or automatic flooding of the area of possible ignition with inert gas, could produce a cheaper and more reliable protection system. One major limitation inherent in these instruments is that they measure the gas concentration of the atmosphere, only immediately adjacent to the sampling point. When they are used to monitor ignition sources or possible leakage points where the geography is well-defined, this can be partly overcome by distributing a number of sampling heads around the source. However, for monitoring leakage of flammable gases, especially where there are a number of possible sites for leakage or where the probable leakage sites are unknown, it is necessary to have some knowledge of the concentration distribution of the gas in space. This can be done using a range of point-measuring devices distributed at suitable points in accordance with the geography of the site. The development of remote scanning systems would however be a better long-term solution to the problem.

In principle, the most convenient way of measuring oxygen concentration is by reduction of the oxygen in a suitable electrochemical cell. Such cells can be produced having geometries and constructions which are rugged, simple and reliable, and indeed a wide range of such instruments with rapid response to changes in oxygen concentration have been produced. Since low oxygen concentrations are a hazard only to the individual at risk and occur relatively infrequently, it is more sensible to mount a continuously operating monitor on the person concerned. Small highly portable instruments have been developed for this purpose, giving rapid warning of a decrease in oxygen concentration to enable the individual to move to a safer position. Since in general the situations where oxygen deficiencies are likely to occur are known, the same instruments can be moved into these areas before the person enters, to check that it is safe. Hence, there is little requirement for large multipoint monitoring systems or for scanning systems for measuring oxygen deficiencies.

The major problem in monitoring toxic gases is that there is no universal physical property of such gases that is related to their hazard in the same way that heats of oxidation are related to LELs for flammable gases. Hence, a wide variety of measuring techniques has been used for measuring toxic gases at concentrations which present an acute risk. Electrochemical techniques are probably the most widely used for the continuous operation of toxic gas alarms which can be carried on the person. Cells similar in construction to those used for oxygen deficiency monitoring but using different electrodes, cell potentials and electrolytes for different gases are increasingly used for measurement of toxic gases at these relatively high concentrations. Such cells can form the basis of small portable instruments which can be carried near the breathing zone of a worker and give an audible alarm when the particular toxic gas in question is present. However, such devices are not generally specific to a particular gas and false alarms may result from the presence of other relatively non-toxic gases.

The most cost-effective way of producing flexible instruments is to use a measuring principle which is sensitive to a wide range of toxic gases but can be made specific to a particular gas by minor changes in operating parameters such as voltage, spectral frequency, etc. This has also led to the development of chemically impregnated paper tape devices; the advantage of these is that the colorimetric methods already developed for a wide range of gases can be used by impregnating paper, or another suitable support material, with the appropriate chemicals and observing the colour changes produced on exposure to the gas.

Many of the instrumental methods are not yet sensitive enough to measure toxic gas concentrations at the levels at which they represent a chronic exposure hazard. The most convenient way of measuring exposure to these levels is by mounting the measuring device on the person at risk. Problems of portability and non-interference with the working practice of the individual has led to the development of small 'air movers' pulling the atmosphere through an absorbent material, which concentrates the vapour for subsequent analyses by laboratory techniques. The adsorbents most widely used at present are small tubes containing charcoal or porous polymeric materials. A significant development in this area is the recent appearance of samplers which utilize molecular diffusion. This technique enables very small convenience badge or tube-type samplers to be used. The low cost of these devices should encourage their widespread use and could result in the continuous routine monitoring of the exposure to vapour of a worker during a shift in the same way that radiation badges are used for monitoring people at risk in the nuclear and related industries.

The continuous monitoring of background concentrations of toxic gases at fixed points in the workplace is generally carried out using modified versions of laboratory instruments, such as gas chromatographs. Gas is drawn from various sampling points through an appropriate tube system and samples from each point are presented in turn for analysis by the instrument. Fixed-point monitoring systems are under development, using the conductivity changes produced in solids by adsorption of the toxic gas on the surface and in time these may well produce monitoring systems similar to those now used for flammable gases. However, perhaps the most widely applicable techniques for toxic gas monitoring are spectroscopic. The combination of infra-red spectroscopic techniques with microprocessor handling of data could result in generally applicable instruments which could easily be adapted for different situations. Increasing interest is also being shown in the use of microwave spectroscopic techniques which have high sensitivity and selectivity.

It will be obvious that most techniques presently used for measuring hazardous gases measure the concentration over a limited amount of time and space. The relationship between these measurements and the degree of hazard for flammable gases, is a problem involving fluid mechanics and statistics. Similarly, where legislation requires that the time-weighted average exposure of a worker be below a certain value, the relationship between the measurement taken at one time and one point, or a number of such measurements, involves a knowledge of the statistical distribution of concentrations in the workplace in time and space so that the probabilities of a man being exposed to a certain dose of gas can be estimated. It is generally realised that it is not satisfactory to take a few measurements in a working atmosphere and from them immediately say whether or not the atmosphere is safe or dangerous. Only the probabilities of safety or danger can be determined. In the development of gas detectors this must be borne in mind, so that the tehniques used reduce the standard deviations in these estimations. In this respect, continuous personal monitoring for toxic gases over the requisite time period is to be encouraged.

All the monitoring methods which have been outlined are relative methods of measurement. That is, they require calibration to produce an accurate measure of the gas concentration. The preparation of standard gas atmospheres, especially over extended periods, for the calibration of instruments, is a field where problems are often underestimated by those who are not familiar with it. The ever-decreasing levels at which concentrations need to be measured and the high chemical activity of many toxic vapours make the preparation of standard gas mixtures a field in its own right. No one technique can be applied to all vapours and it is essential to know the extent to which the gas is likely, for example, to react with other components in the mixture or to be adsorbed on the walls of the vessel, if reliable results are to be obtained.

Monitoring of hazardous gases is a relatively new field and one in which rapid developments are taking place. It provides a direct answer to a real health and safety problem in industry and, when coupled with automatic remedial methods, may well provide the most reliable and cost-effective answer.

1 History and law

J. H. Burgoyne

The history of the detection and measurement of dangerous gases in air is long and rather detailed. Only an outline can be attempted here. Indeed, it may be judged, depending on personal viewpoint, that what is presented is a dipping into history rather than a balanced summary.

The law presented is that of the United Kingdom with a short review of the situation in the United States of America. In going further, it would be difficult to know where to stop. What is presented does however give some idea of the extent to which the detection and measurement of hazardous gases is becoming a statutory requirement.

1 History of concern with dangerous contaminations of the workplace atmosphere

Man's earliest pre-occupation was the quest for food. It may be speculated that his first occupational injuries were bruises and abrasions, together with eye injuries as he chipped out his primitive rock weapons. Perhaps he contracted anthrax as he skinned his game.

Specialist occupations appeared as agrarian societies developed, the first being the winning of metals from their ores. As civilisation developed, the system of slavery was evolved to cope with laborious and unpleasant tasks and there was little concern with what we would now call occupational medicine. Indeed stratification into citizens and workers became pronounced and the writings of Hippocrates, the doyen of medicine, are almost exclusively concerned with the health of the former.

One of the earliest known writings that refers to occupational medicine is the *Historia Naturalis* of Pliny the Elder, which dates from the first century A.D. He recommends that a protective mask consisting of a crude bladder be placed over the face of a worker to protect him from the inhalation of dust during the grinding of mercury (cinnabar) and of fumes from the smelting of lead.

The Dark Ages, as the name describes, saw regress not progress, and the Middle Ages were characterised by another social stratification into landlords and serfs. As the Renaissance approached, peasants drifted to the towns and merchants, artisans and labourers emerged, bound together by the guild system. Unfortunately, owing to ignorance of the rudiments of public hygiene, epidemic disease was rampant. Only gradually and intermittently does occupational disease become discernible, still in the fields of mining and metallurgy. Georgius Agricola's classic *De Re Metallica*, published in 1556, contains a section describing the accidents and diseases to which miners were heir and recognising the benefits of ventilation and fresh air. The true pioneer of occupational health and hygiene was however Bernardino Ramazzini who in his *De Morbis Artificum Diatribe* (1700) described the occupational diseases of his time and proposed remedies that were to be ignored for more than two centuries.

In the late eighteenth century, the Industrial Revolution greatly broadened the range of occupational hazards. A treatise on the subject by Charles Thackrah, appearing shortly after 1800, was the first of its kind to be published in England. The Pneumatic Institution, founded a little earlier by Thomas Beddoes for the study of inhalation therapy in the treatment of respiratory disease, had Humphry Davy as its superintendent.

The nineteenth century saw concern for the health of the worker at a low ebb in relation to the growing size of the problem. The trade union movement gave a voice to the workers and employers began to realise that workers' welfare was in their own economic interest. Genuine humanitarianism and utopian socialism created a turning-point and in the early years of the twentieth century modern occupational medicine was born. At the same time advances in science and technology opened the way to more effective reduction of exposures and the physical protection of the worker. English factory and health laws led the way, but co-ordination in practice of the skills of the chemist and the engineer with those of the medical man has come slowly. As late as 1942, E. R. A. Merewether, Senior Medical Inspector of Factories in England, thought it necessary to say to an American Public Health Association audience that 'the doctor no longer knows everything about everything' and to concede American leadership in industrial hygiene engineering.

The particular contribution of the industrial hygiene chemist is the detection and measurement of toxic contaminants of the working atmosphere. Co-operation is necessary with the field industrial hygienist in obtaining the appropriate air samples from the workplace, or in selecting appropriate locations for *in-situ* measurements.

The attitude to exposure has changed remarkably during the present century. As the earlier legislation of the period shows, entry into doubtful atmospheres was at first either avoided, or undertaken with the protection of breathing apparatus. Little reliance was placed upon sophisticated methods of analysis either for warning or for supervision. Analytical techniques received much impetus from interest in, and legislation upon, atmospheric pollution control in the mid-century and this has facilitated the sensitive monitoring and supervision of working atmospheres, particularly in relation to so-called 'threshold limit values'. International disagreement on actual figures and a continual downward drift in recommended values has encouraged the view, especially in international trade union circles, that no detectable exposure to a toxic substance is acceptable. Should this view prevail, emphasis would shift, more than ever, to totally-sealed and remotely-controlled process systems and the industrial hygiene chemist could be required to provide not so much methods of measurement as highly sensitive techniques of detection.

2 History of the detection and measurement of toxic substances in air

From early times in industrial history, birds and small animals, such as mice, have been used to monitor the respiratory condition of the working atmosphere. They have also been used for purposes of research longer than might be thought. An essay by C. T. Coathupe[1], medical student turned industrialist, published in 1838, examines the state of knowledge on suffocation due to lack of ventilation and oxygen depletion by combustion processes. The writer remarks too on the fallacy of applying the results of tests with toxic substances on small creatures in confined spaces to men in open workplaces and roundly declares that his foundrymen, despite regular exposure to copious arsenical fumes, are healthy, long-lived and 'with numerous progeny'. He concludes robustly, in terms that have a certain relevance to present-day problems: 'If manufacturers are to be crushed by the results of experiments performed upon birds and rabbits, in boxes, merely because an unwanted odour may happen to annoy the fastidious olfactory organs of some idle independent residing within their vicinity – while the broad fact is that multitudes are obtaining their support in health and ease, who are more immediately connected with the influence of the alleged nuisance – how can we expect science to advance in the useful departments of arts and manufactures?'

In the nineteenth century, the analysis of air followed the methods of the early chemists. Oxygen had been measured by Priestley and Cavendish in the late eighteenth century, using nitric oxide as reagent, but their inability to prepare this gas in a pure state limited the accuracy of measurement. By 1890, according to Wanklyn and Cooper,[2] this difficulty had been overcome, the nitrogen dioxide formed being absorbed in water to allow the relevant contraction to be measured. The method of explosion with hydrogen, 'esteemed by Bunsen and Regnault' was also available, and the method of absorption in alkaline pyrogallol had come into use.

The measurement of carbon dioxide by absorption in lime water or barytes was the simplest and possibly the earliest measurement of an air contaminant to be made, but Bunsen in the mid-nineteenth century was able to do a full analysis of illuminating gas using absorption reagents and explosion analysis, with nitrogen the final residue.

The state of the art prior to the First World War can be seen clearly in the first edition of J. S. Haldane's *Methods of Air Analysis* (1912).[3] Portable apparatus for the measurement of carbon dioxide, methane and carbon monoxide in the air of mines 'worked with naked lights', and for 'routine methane estimation' is described. A test is given for the approximate estimation of oxygen by means of a lighted taper lowered into a jar of sampled air and extinguished by the combined effects of oxygen concentration, convection and suffocation (Fig. 1.1). Details are recorded of the estimation of methane from flame-caps (see later) and of carbon monoxide by bird and small animal tests, as well as by a blood drop test. Other analyses recorded are for hydrogen sulphide with lead acetate, for nitrous fumes with starch–iodide, and for phosphine with silver nitrate, but these were in use in the early nineteenth century, as indicated by Coathupe.[1]

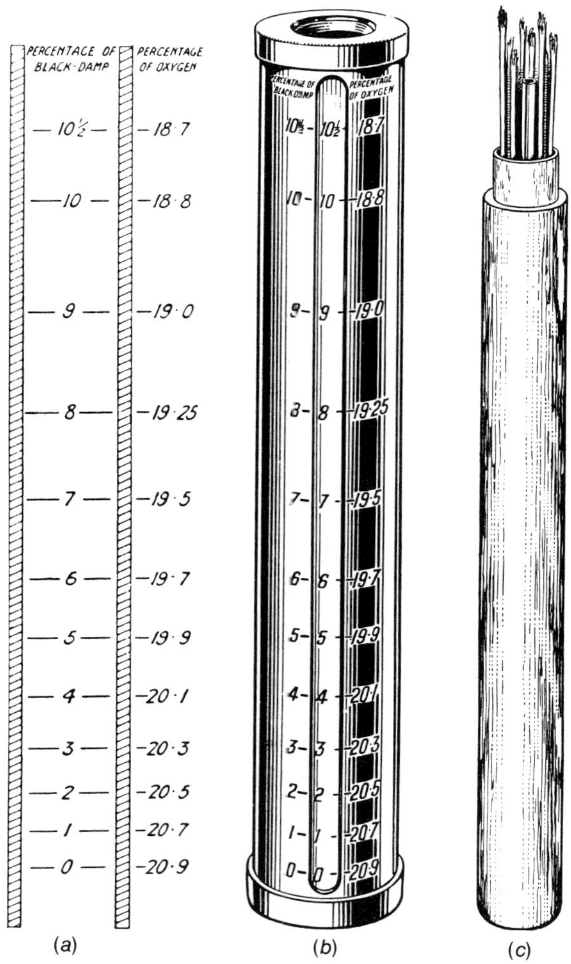

Fig. 1.1 Apparatus for determining oxygen by flame test (Haldane). (*a*) Section of glass tube; (*b*) tube in aluminium case; (*c*) inner wooden tube with tapers and brass holder. By permission Charles Griffin and Company Ltd.

After the First World War, the Haldane type of portable apparatus continued in use for the analysis of mine and other industrial air (Fig. 1.2). The use of air-aspirated dry reagent detector tubes, giving progressive colour changes for a number of specific toxic gases and vapours was pioneered by the newly-formed Department of Scientific and Industrial Research and published in a series of booklets entitled *Methods for the Detection of Toxic Substances in Air.* These are perpetuated and extended in the current series under the same title published by the Health and Safety Executive and the method has been extensively developed commercially, especially by Drager.

Other Post-Second World War advances have been concerned with the

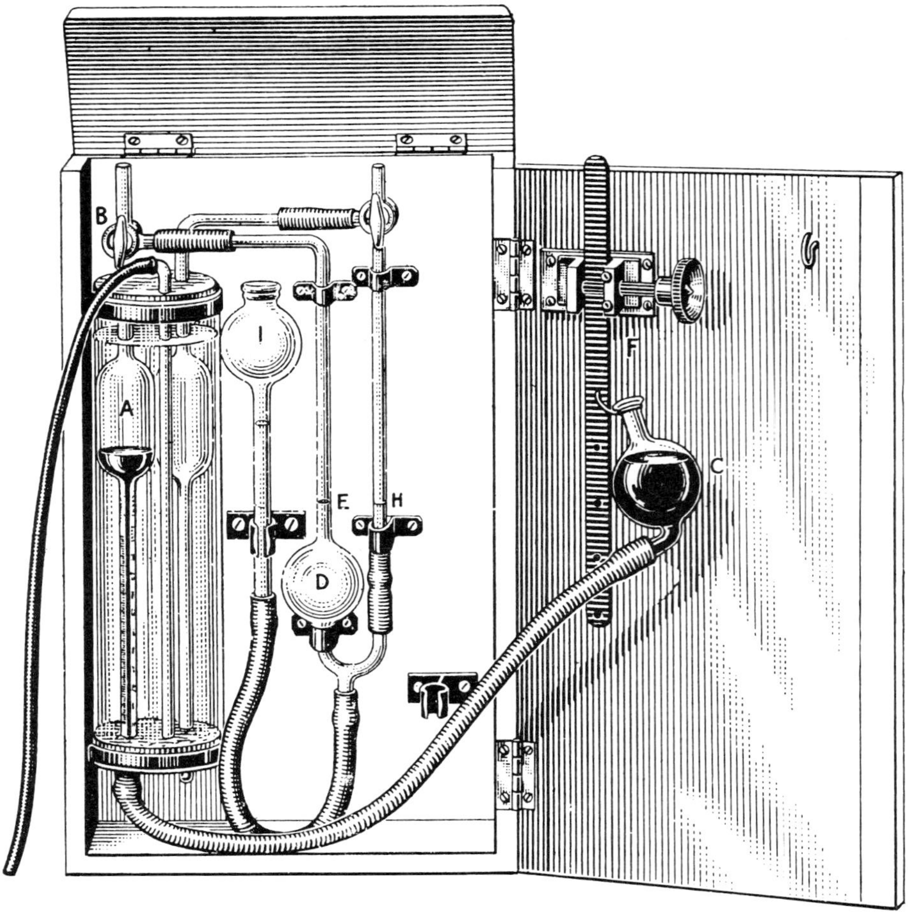

Fig. 1.2 Portable apparatus for determining small percentages of carbon dioxide in air (Haldane). By permission Charles Griffin and Company Ltd.

development of increasingly sensitive chemical and physical methods of analysis and their translation into rapidly-responding instrumental techniques of detection, measurement and recording. These form part of the contemporary scene, which is considered in Chapters 4 and 5.

3 History of the detection and measurement of flammable gases in air

These seems little doubt that industrial explosions due to flammable gas in the working atmosphere were first encountered predominantly in coal mining. The release of 'firedamp' – essentially methane – from the coal measures being worked must have bedevilled the industry from the start. From reference in the earlier

Fig. 1.3 The First Fireman. An impression of a 'fireman', protected by a woollen gown, burning out firedamp, accumulating under the roof of a mine working, using a long taper (Photo. Science Museum, London).

numbers of the *Philosophical Transactions of the Royal Society* it appears that the danger from firedamp was well-known in the late seventeenth century (Fig. 1.3). The risk of ignition by a candle or sparks, leading to explosion, was appreciated. At the same time, the indication of the presence of non-explosive levels of gas in the mine atmosphere, furnished by the appearance of a candle flame taken into it, seems to have been understood, and the earliest form of 'methane detector' took the form of a lighted candle. The following account of its use was given later by Mr John Buddle in a letter addressed on 18 October, 1813 from 'Wall's End' Colliery to the Society in Sunderland for Preventing Accidents in Coal Mines. He pointed out that every quantity of gas, from the slightest mixture to the firing point, could be discerned by the experienced collier from observation of the appearance of the spire upon the top of the candle and continued:

> The common pit candles vary in size, but those generally used are forty-five to the pound; the wick is of cotton, and the candle made of ox or sheep tallow; but clean ox tallow is the best.
>
> The mode of trying the candle, as it is called, to ascertain the mixture of inflammable gas, is as follows:
>
> In the first place, the candle, called by the colliers the 'low', is trimmed, that is, the liquid fat is wiped off, the wick snuffed short, and carefully cleaned of red embers, so that the flame may burn as purely as possible.
>
> The candle being thus prepared, is holden between the fingers and thumb of the one hand, and the palm of the other hand is placed between the eye of the observer and the flame, so that nothing but the spire of the flame can be seen, as it gradually towers above the upper margin of the hand. The observation is generally commenced near the floor of the mine, and the light and hand are gently raised upwards, till the true state of the circulating current be ascertained.

> The first indication of the presence of inflammable air is a slight tinge of blue, or bluish grey colour, shooting up from the top of the spire of the candle, and terminating in a fine extended point. This spire increases in size, and receives a deeper tinge of blue as it rises through an increased proportion of inflammable gas, till it reaches the firing point; but the experienced collier knows accurately enough all the gradations of 'shew' (as it is called) upon the candle, and is very rarely fired upon, excepting in cases of sudden discharges of inflammable gas.
>
> This shew upon the top of the candle varies very much according to the length of run, or distance, which the current of air has passed through, before it is mixed with the inflammable gas. The shorter the run of current of air, before it is mixed with the inflammable gas, the less will be the shew upon the candle when at the firing point, and vice versa.

Between 1730 and 1750, the flint-and-steel mill was developed by Mr Carlisie Spedding of Whitehaven. This gave light by the continuous generation of friction sparks, accompanied by a somewhat reduced, but by no means non-existent, risk of ignition. The presence of non-explosive concentrations of firedamp in the air was indicated by a growth in size and luminosity of the sparks.

Although the 'candle watchers' exercised considerable skill in assessing the dangers to their fellow workers, the method of detection was itself highly dangerous. Even the Spedding Mill was seen to be far from safe. By the beginning of the nineteenth century coal mining was expanding rapidly in response to the demands of the Industrial Revolution, and this brought an increasing frequency of alarming explosions. A safe means of lighting and gas detection became the urgent need. So began the long and complex history of the safety-lamp, recounted by R. L. Galloway in his *Annals of Coal Mining* and by Hardwick and O'Shea in the *Transactions of the Institution of Mining Engineers for 1915–1916.*[4] Many other accounts have been given.[5] Only an outline need be attempted here.

At the beginning of 1812, Dr W. R. Clanny, a physician living at Bishopwearmouth, near Sunderland, began the development of a series of oil lamps in which the air for combustion was forced in through a water-trap and the products passed out likewise, or through a tall chimney tapered to a fine point. In this way he sought – with some success – to isolate the flame inside from the possibly explosive atmosphere without. His first account was contributed to the Royal Society on 20 May 1813.[6]

About a year before, on 25 May 1812, there occurred at Felling Colliery, near Jarrow, an explosion that cost ninety-two lives. As a result, a society was formed in Sunderland to press for an intensive study of the causes and means of prevention of such disasters. Arising from this, in the summer of 1815, Dr Gray, the rector of Bishopwearmouth, secured the interest of Sir Humphry Davy, then Professor of Chemistry at the Royal Institution in London.

The first basis of this approach was as to the possibility of preventing explosions by chemical inhibition. Davy seems to have rejected this idea quite soon however, and to have embarked on a short programme of research into the combustion of methane that led him to the version of the safety lamp with which his name is associated. In this, the flame of the oil lamp was isolated from the ambient atmosphere by being entirely surrounded by fine wire gauze.

His essential conclusions were presented to the Royal Society in a paper read on 9 November 1815.[7] In January 1816 the first Davy lamps were tried out successfully in some gassy parts of Hebburn Colliery.

In 1815 also, George Stephenson, later to become an eminent railway engineer, was enginewright at Killingworth Colliery and turned his attention to the development of a safety lamp. More or less simultaneously with Davy, he produced lamps in which the flame compartment was separated from the surrounding atmosphere by perforated plate, small holes and fine gauzes. Stephenson did not immediately publish his findings and much dispute as to priority of invention arose subsequently. History has awarded the verdict to Davy, and perhaps justly so, if only because his researches led not only to the safety lamp, but also to a considerable increase in knowledge of the science of combustion.

As Buddle had made clear in the extract quoted earlier, there was a considerable understanding in the days of the 'candle watchers' of the effect of the presence of flammable gas in the mine atmosphere in elongating the candle flame; also of the correspondence between the character and extent of this elongation and the concentration of flammable gas in the atmosphere, leading towards the condition that would result in a general explosion.

The safety lamp could be used as a gas detector in the same way as the candle, though observation was more difficult owing to the surrounding screen. There was prejudice against its use on this ground as well as on grounds of convenience and efficiency. The same Buddle said to a Parliamentary Committee in about 1835: 'I consider it a very great protection, the using of candles. It keeps every man upon the alert: we never can by the use of the Davy lamp judge the quantity of air passing so well as we can do with the candle.'

A further function of the candle flame, and hence of the safety-lamp also, was indicating a deficiency of oxygen in the working atmosphere. This results in a falling off in the luminosity of the flame. According to Forbes and Grove,[8] the illumination from a properly-set safety-lamp decreases progressively from 100% in the normal oxygen content of the atmosphere of 20.94% by volume to 12% at 18.30% by volume and is extinguished at about 16.25% by volume. Its sensitivity to the presence of methane falls correspondingly and it is considered to be ineffective in this respect in atmospheres containing less than about 19% by volume of oxygen. With sufficient oxygen in the atmosphere, the safety lamp can, it is claimed, be used without undue difficulty to detect methane from about 1% by volume when the first signs of flame elongation are apparent, to about 5% by volume when the flame is extinguished owing to extreme expansion. The lower explosive limit of pure methane in air is recorded as 5.3% by volume. To experienced eyes therefore, the safety-lamp gives a visual impression of the condition of the mine atmosphere not easily obtainable in any other way. For this reason, its use, in improved forms, as a methane detector has continued to the present time. As a means of illumination it has of course been superseded long since by electrically-powered lamps designed for use in potentially flammable atmospheres.

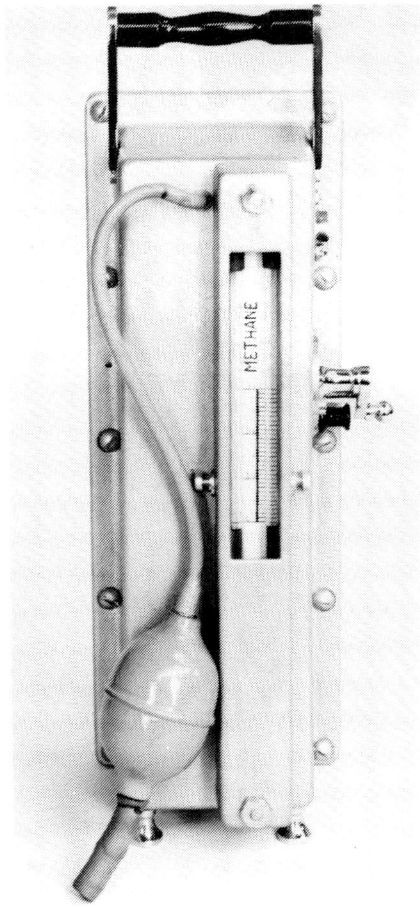

Fig. 1.4 McLuckie's firedamp detector (Crown Copyright. Science Museum, London).

Variations in and improvements to the Davy lamp made or suggested during the 160 or so years since its invention have been most numerous. Much detail is to be found in the reviews already quoted. Developments affecting its use as a gas detector have included: arranging the air inlet for close application to the roof of a mine roadway where the concentration of methane (which is lighter than air) is likely to be highest (Gray lamp, 1868); the use of scales to measure the flame extension indicating the concentration of methane in the air, sometimes with a lamp fuel alternative to oil (Clowes lamp using hydrogen, Stokes and Peiler lamp using methanol and Chesnau lamp using methanol with additives); the use of bimetallic and other temperature-measuring devices to register flame height and possibly to actuate an alarm (Koehler lamp); and, most ingenious, a sound tube to produce an audible warning at a chosen flame height (Fleissner singing flame lamp).

In the final outcome, it seems that, although claims have been made for the detection of as little as 0.25% of methane in air with the flame lamp, estimation of

Fig. 1.5 Ringrose firedamp detector (Crown Copyright. Science Museum, London).

firedamp in this way is a subjective process, especially in concentrations of less than 2%.[9]

Alternatives to the flame safety lamp as a gas detector took some time to emerge. The main incentive would be the desire to produce a more quantitative instrument. The first achievement in this direction was probably that described by Monnier in 1881.[10] A sample of air containing flammable gas was burnt on an electrically-heated filament in a sealed chamber. After allowing time for cooling and condensation of water, the reduction in pressure was observed. The principle was applied in the McLuckie Methanometer[11] (Fig. 1.4), which Hartwell[5f] referred to in 1950 as being the instrument most widely-used in the mines of the United Kingdom after the flame safety lamp. The Burrell detector was a version used in the United States of America.

A quicker mode of operation was achieved by burning the mixture continuously on a filament situated in a porous-walled chamber through which the mixture flowed.

Fig. 1.6 Liveing flammable gas detector (Crown Copyright. Science Museum, London).

A reduction in pressure proportional to the concentration of flammable gas passing through the chamber was observed due mainly to the condensation of the water formed, and operation was continuous. The method was first described by Crossland in 1928[12] and found application in the Ringrose Methanometer (Fig. 1.5). This is equipped with a signal lamp actuated by the pressure difference produced and it has been chiefly used in the UK as a gas alarm. An instrument depending for its action upon the relative rates of diffusion of the unburnt methane and air through the porous chamber walls and the resultant pressure change in the chamber was the German Neufeldt and Kuhnke methane detector.

The present era of flammable gas detecting and measuring instruments dates back to the attempt in 1880 by Liveing[13] to estimate the flammable gas content of air by measuring the heat generated when the mixture reacts on a heated filament (Fig. 1.6). The fact that the heats of combustion of most lower flammability limit mixtures are very similar to one another renders this a singularly appropriate principle on which to base the design of an instrument to measure flammable gas content, since, as a first approximation, the response to any given fraction of the lower limit concentration is independent of the nature of the flammable gas. The application to situations other than mining, where a variety of flammable gases other than methane may appear, is therefore facilitated.

Instruments of this kind rely on the measurement of the change of electrical resistance of the heated filament on which the gas is burnt catalytically. The filament is included in one arm of a Wheatstone bridge and the out-of-balance current may be taken as a measure of the flammable gas content of the sampled air. This type of instrument may in principle be used as a gas detector or for continuous measurement and, again in principle, it may be utilised to give an alarm or to effect normal process or emergency control. Since the heated filament could provide a source of ignition if introduced into a flammable gas–air mixture it must be protected, by flame arresters or otherwise, to render it safe for use in all circumstances. Technical improvements, such as the replacement of filaments by pellistors, to give added stability, reliability and durability to such instruments, are a feature of the contemporary scene and will be discussed in Chapter 2, as will also be the use of semi-conductor gas-sensing elements.

4 Odorisation of distributed gas

The odorisation of hazardous gases has a particular role in the detection and location of escapes. Traditional town gas, manufactured for well over a century from coal and coke, had a natural and distinctive odour attributable to a small content of organic sulphur compounds, aromatic and polycyclic hydrocarbons. The increasing use, during the present century, of petroleum products as a source of distributed gas led to a loss of natural odour and the need for added artificial odorant. Tetrahydrothiophen (THT) was widely adopted and a scale of odorosity developed based on the relative volume of air required to dilute the odorised gas to the limit of detection.
This odorant proved to be less effective with North Sea gas when this was first introduced into Britain by British Gas from the West Sole field at Easington in the mid-1960s. It was concluded that the heavy hydrocarbons which the gas then contained de-sensitised the nose to the particular odour of THT. A new odorant was therefore sought, with the result that a mixture of 90% dimethyl sulphide and 10% mixed mercaptans was adopted.

When gas from the major Leman Bank Field was landed at Bacton in the later 1960s, this was found to be odorous, owing to a small content (some 10 p.p.m.) of organic sulphur compounds, although judged to be lacking in 'impact'. This latter quality was imparted by a very small addition of tertiary butyl mercaptan (TBM) and eventually an odorant blend was adopted consisting of the major sulphur compounds present in Leman Bank gas (diethyl sulphide and ethyl mercaptan) and TBM. This formulation, known as Odorant BE, has fulfilled the role of universal odorant for distributed gas in Great Britain since 1973. A scale of odour intensity has been adopted and skilled observers, known as 'rhino-analysts', have been trained. Much research has been done by British Gas on the effect of gas distribution and handling, and leakage through the soil, on gas odorisation, and also upon olfactory fatigue, and further developments are likely.

In the USA and Canada, odorants similar to BE are applied to distributed natural gas. In Europe only THT is used up to the present, although problems of matching odorisation of gas from different sources have begun to appear.

5 United Kingdom law requiring the detection and measurement of hazardous gases

The Health and Safety at Work etc. Act 1974 makes 'provision for securing the health, safety and welfare of persons at work, for protecting others against risks to health or safety in connection with the activities of persons at work' and 'for controlling certain emissions into the atmosphere'. The implication of the control of the harmful content of the atmosphere to which those at work, and others, may be exposed is clear.

The Act 'adopts' a number of pre-existing enactments of safety law as 'relevant statutory provisions' but provides for their progressive replacement by 'a system of regulations and approved codes of practice' operating in combination with the other provisions of the Act and designed to maintain or improve the standards of health, safety and welfare established by or under the existing enactments.

5.1 The working environment

Particularly relevant in connection with working atmospheres is the Factories Act 1961, which in the more restricted context of a 'factory' makes a number of provisions for the control of potentially dangerous working environments.

(1) Section 30 deals with work in confined spaces in which 'dangerous fumes are liable to be present to such an extent as to involve risk of persons being overcome thereby', or 'the proportion of air is liable to have been substantially reduced'. Such spaces may not be entered by anyone not wearing suitable breathing apparatus unless sources of continuing contamination have been isolated or removed, adequate continuing ventilation has been provided and the atmosphere has been tested for dangerous fumes and the adequacy of oxygen for respiration.

(2) Section 31 deals with a 'plant, tank or vessel which contains or has contained any explosive or flammable substance' and forbids hot work upon it until 'all practicable steps have been taken to remove the substance and any fumes arising from it, or to render them non-explosive or non-inflammable'. No specific reference is made to testing of the contained atmosphere but clearly this would be included in 'all practicable steps' in the present state of the art.

(3) Section 63 provides that where 'in connection with any process carried on, there is given off any dust or fume or other impurity of such a character and to such an extent as to be likely to be injurious or offensive to the persons employed, or any substantial quantity of dust of any kind, all practicable measures shall be taken to protect the persons employed against inhalation of the dust or fume or other impurity and to prevent its accumulating in any workroom'. Reference is then made to local extract ventilation. Once more testing of the atmosphere of the workroom is not specifically mentioned, but there is no doubt that it is now entailed in 'all practicable measures'. Indeed, where reliable methods of continuous monitoring are available, these could be deemed to be included as well.

The intended application of these provisions of the Factories Act through the Health and Safety at Work etc. Act is clearly to all places of work. Some specific applications, currently effective under the 1961 Factories Act and its predecessors, should be mentioned.

The Chemical Works Regulations 1922, originally made under the Factory and Workshop Act 1901, apply to certain named manufactures and associated processes carried on in chemical works. There is a requirement for enclosure and local extract ventilation in connection with certain fume- and dust-producing operations; for the control of certain potential sources of ignition where flammable gases, etc. may arise; for properly maintained breathing apparatus and other emergency equipment where dangerous gas or fume may escape; for the isolation of sources of gas or fume from confined spaces that are to be entered, and the provision and use of properly maintained breathing apparatus and a life-belt. No reference is made to the testing of atmospheres. It may be doubted whether reliable equipment of a sufficiently wide range of application was available at the date of these regulations.

The Shipbuilding and Ship-repairing Regulations 1960, made under the Factories Acts of 1937 and 1948, refer to 'the construction and repair of ships and vessels in a hard or dry dock and in the construction and repair of ships (but not of vessels other than ships) in a harbour or wet dock'.

It is required that in confined work places adequate ventilation (with air, not oxygen) be provided (Reg. 48), that inflammable atmospheres be precluded by ventilation or other means (Reg. 49), that oxygen deficiency be guarded against by ventilation and testing or the use of breathing apparatus (Reg. 50), and that safety from the presence of dangerous fumes be achieved by isolation or removal of the source, by ventilation and testing (the safe period thereafter being specified), or by the use of breathing apparatus (Reg. 51).

Oil-carrying vessels are subject to further restrictions concerning the introduction of naked lights and other potential sources of ignition except where period certificates of sufficient freedom from inflammable vapour, etc. have been given, or certified as unnecessary, by a competent analyst (Reg. 59). The extension of this provision to bilges, shaft tunnels, pumprooms, lamprooms and other compartments and spaces is emphasised (Reg. 65). A specific reference is made (Reg. 60) to oil-tanks that have contained low flash-point oil (below 73 °F Abel closed test), to which entry is forbidden unless 'an analyst has certified that the atmosphere is sufficiently free from inflammable mixture'. The certificate must specify a period of validity (Reg. 61). The entry of the analyst for the purpose of certification is not precluded, but clearly sampling from outside would be preferred.

In the case of an oil tank, cleaning and ventilation must precede testing and certification (Reg. 63). Period certificates are invalidated by changes affecting the isolation of the source of flammable material (Reg. 64).

Some exceptions and relaxations are made in connection with minor repairs to tanks that have carried only high flash-point oils (above 150 °F, Pensky-Martens closed-cup) or liquid methane, propane or butane at atmospheric pressure (Reg. 66).

The Construction (General Provisions) Regulations 1961, also made under the Factories Acts of 1937 and 1948, relate to building operations and works of engineering construction. They require that all reasonably practicable measures be taken to prevent the inhalation of evolved dust and fumes by adequate ventilation, by the provision and use of suitable respirators, or otherwise (Reg. 20). More specifically, 'every working place in any excavation, pit, hole, adit, tunnel, shaft, caisson or other enclosed or confined space' and every approach to any such working place shall be adequately ventilated to make it safe for respiration and where any danger of an unsafe atmosphere is apprehended entry is forbidden until, following suitable testing, a competent person 'is satisfied that the working place or approach is, for the time being, free from the danger of a person being overcome by poisoning or asphyxiation' (Reg. 21). The Regulations do not require the testing of the atmosphere of any and every confined workplace irrespective of any apprehension of contamination (Reg. 22).

The Health and Safety at Work Act is administered by the Health and Safety Executive (HSE) who indicate their practical interpretation of certain aspects of regulations in their official publications. Some such views on flammable gas detectors are to be found in Guidance Note CS/1 (December 1976) entitled *Industrial use of flammable gas detectors.* This refers chiefly to detectors of the catalytic element type.

Some typical situations in which dangerous concentrations of flammable gas or vapour can arise and in which fires and explosions have occurred are mentioned. These include the removal of flammable materials from confined spaces or enclosures prior to entry; the evaporation of solvents in drying processes; the use of flammable atmospheres in heat treatment furnaces; the application of coatings in flammable solvents; the manufacture of gases; the manufacture and mixing of flammable liquids, such as solvent paints and inks; the charging of batteries; hydrogenation processes; the combustion of gas or oil; and solvent extraction processes. The cleaning, repair and demolition of tanks for storing flammable liquids has also led to accidents and is considered in a separate publication – Technical Data Note 18 – due for re-publication.

In these and other demonstrably dangerous situations, the use of flammable gas detectors and monitors is of considerable, possibly vital, assistance in avoiding the actuality of danger. An important reservation is of course that the detector itself shall not be a source of igition of a flammable atmosphere. This is a requirement of Regulation 27 of the Electricity (Factories Act) Special Regulations 1908 and 1944. In practice this means that instruments must be certified for the gas (or apparatus) group (or class) appropriate to the gas being measured, by an established authority, as conforming to an acceptable standard of explosion protection. An appropriate standard is that of the British Approvals Service for Electrical Equipment in Flammable Atmospheres (BASEEFA) – Health and Safety Standard SFA 3007: 1972, *Instruments for Measuring Gas Concentration.* All flammable gas detectors certified in the UK up to 30 September 1969 are listed in the Department of Employment's Publication entitled *Intrinsically Safe and 'Approved' Electrical Apparatus* (form F931). Those certified in the UK after that date are listed in the current issue of the BASEEFA List of Certified Equipment. Before the BASEEFA Standard was issued, flammable gas detectors were certified in the UK as intrinsically safe to British Standard 1259.

An important precaution in practice is that flammable gas detectors should be calibrated before use with the aid of a standard atmosphere (see Chapter 7).

Serious accidents have occurred, and continue to occur, whilst work is being done inside confined spaces. The cause of accidents may be the accumulation and ignition of flammable gases or vapours, but is more commonly the accumulation of toxic gases or vapours, or the existence of a deficiency of oxygen. The practical control of these hazards is considered in the HSE Guidance Note GS5 (July 1977) entitled *Entry into confined spaces.* Examples are listed as: gas or vapour remaining from a process previously carried on in the space; gas or vapour which enters the space due to ineffective isolation from sources elsewhere; fumes emitted when sludge or deposits are disturbed; fumes produced by such operations in the space as welding, painting and lining; products of fuel combustion. A special risk arises when oxygen is released and enriches the enclosed atmosphere, so rendering materials unusually ignitable or combustible. Oxygen deficiency can arise from inert gas purging or other air displacement mechanisms, or by the consumption of oxygen through combustion or oxidation (including corrosion) processes.

The preparation of confined spaces for entry involves an organised procedure of isolation, cleaning and purging, followed by atmospheric testing and monitoring and finally certification for entry, with an emergency control plan in hand. In this procedure the use of instruments or methods to detect and measure toxic contaminants and oxygen sufficiency, either on a periodical or a continuous basis, plays an essential part.

The method of test or measurement to be employed depends of course on the nature of the toxic substance or substances present. Officially favoured methods include those described in the series of twenty-six individual booklets: *Methods for the detection of toxic substances in air,* obtainable from HMSO Government Bookshops and elsewhere. Calibrated chemical detector tubes and portable oxygen meters are also mentioned. Flammable gas detectors are not usually suitable for the detection and measurement of toxic gases, being insufficiently sensitive even if they respond. The use of small animals such as mice and canaries to test atmospheres for respirability is not now favoured by HSE.

Official recommendations regarding highest concentrations of airborne materials to which people should be exposed day by day in the working atmosphere are tabulated in HSE Guidance Note EH15/77 *Threshold Limit Values for 1977.* This information is based on a companion publication of the American Conference of Governmental Industrial Hygienists and is updated annually. It is emphasised that threshold limit values (TLVs) do not represent sharp dividing lines between safety and danger and the best practice is to keep contaminants of the working atmosphere as dilute as possible, with the TLV as the highest permissible time-weighted average concentration. Information is also given regarding maximum permissible concentrations for 15-minute exposures, and also for instantaneous exposures ('ceiling values'). About 600 substances are listed.

The Mines and Quarries Act 1954 is a 'relevant statutory provision' under the Health and Safety at Work Act. It imposes a duty in connection with coal and other mines (except in stopped-off parts or waste) to provide adequate ventilation, 'diluting gases that are inflammable or noxious so as to render them harmless', and 'providing air containing a sufficiency of oxygen'. Specifically, the level of carbon dioxide is not to exceed 1.25% by volume and that of oxygen is not to be less than 19% by volume (Section 55). Steps must be taken to minimise emissions from waste or the dangers therefrom.

The Coal and Other Mines (Ventilation) Regulations 1956, made under the

Mines and Quarries Act and varied by Regulations of 1960 and 1966, go into considerable detail regarding practical arrangements to be made in fulfilment of the conditions of the Act. For example, flammable gas in intake airways up to 300 feet from the first working face shall not exceed 0.25% by volume in six samples from the general body of air taken at not less than fourteen days interval (Regulation 2). Other limits are laid down for other places. Sampling and analysis must be carried out by a competent person using approved apparatus. Results must be recorded and the Inspector notified of determinations exceeding 1% by volume of flammable gas. Approved detectors, available in requisite numbers, are to be maintained and tested as directed in the certificate of approval (Regulation 10). The definition of 'competent persons' is set down, as is that of persons certified to use flame safety lamps. Automatic detectors are to be used for flammable gas levels in excess of 0.5% by volume. In the presence of flammable gas, flame safety lamps are to be used in a prescribed manner. No other lamp is to be used in these circumstances (Regulation 10).

The Coal and Other Mines (Fire and Rescue) Regulations 1956 were also made under the Mines and Quarries Act. They require the provision amongst rescue equipment of the historic means of detecting flammable gases and non-respirable atmospheres. The equipment of the Rescue Station shall include six flame safety lamps and the means for training persons to test for flammable gas with such lamps. Also to be provided are cages of small birds for testing for noxious gas (Regulation 23). Except for small mines, the rescue equipment at every mine shall include one flame safety lamp for every man trained in rescue work, and two small birds for testing for noxious gas (Regulation 24).

The Coal and Other Mines (Safety-Lamps and Lighting) Regulations 1956, in Part II, go into some detail regarding the provision, maintenance and use of flame safety-lamps for the detection of firedamp.

The Coal Mines (Firedamp Drainage) Regulations 1960 require that any underground firedamp discharge point be surrounded by a fence at a distance where the level of firedamp in the air does not exceed 2% by volume (Regulation 13).

Safety in offshore installations is regulated by the Mineral Workings (Offshore Installations) Act 1971, extended and amended by the Petroleum and Submarine Pipelines Act 1975. Two statutory instruments made under these Acts have relevance to the detection and measurement of hazardous gases in working atmospheres.

The Offshore Installations (Operational Safety, Health and Welfare) Regulations 1976 require that 'hazardous areas', in which there is likely to be danger of fire or explosion from the ignition of gas, vapour or volatile liquid, shall be identified, and clearly and accurately recorded in the operations manual (Regulation 2). Hot and other work giving rise to a source of ignition and work in places dependent upon ventilation for diluting and dispelling injurious or flammable fumes etc. and providing sufficient oxygen, shall only be carried out under a work permit system and in accordance with written instructions (Regulation 3). Specific mention is made in Schedule 2 of the necessity for written instructions to provide for detection of and protection from sour gases (i.e. containing hydrogen sulphide). Although gas detectors and monitors are not mentioned, their use in implementing the regulations is clearly implied.

The Offshore Installations (Fire-fighting Equipment) Regulations 1978 do, however, specifically require that 'every offshore installation shall be provided with an automatic flammable gas detection system capable of monitoring continuously every part of the installation in which flammable gas may accidentally accumulate' (Regulation 6). Moreover, audible and visible response to any accumulation of flammable gas shall be given at a continuously manned place. In addition, any manned installation shall be provided with two 'portable devices capable of measuring accurately an accumulation of flammable gas'.

In looking through the above extant legislation covering the last 50 to 60 years, with a sense of historical perspective, certain trends connected with the detection and measurement of hazardous gases can be discerned.

(1) Concern came first with the dangers of special enclosed and unventilated situations and only later with the working atmosphere generally.

(2) Analysis of the atmosphere, not at first mentioned specifically, has increasingly become recognised as a tool for controlling the risks.

(3) In occupations where the risk has been found, or thought to be, exceptionally great, stress has been placed at a relatively early stage upon measurements of some precision, as in coal mining long since, and upon continuous monitoring, as in the offshore field recently. Such emphases have undoubtedly provided a stimulus to technical development.

5.2 The public environment

The Health and Safety at Work etc. Act 1974 provides for the protection of the public from dangers arising from work activities and, in Section 5, makes particular reference to the control of harmful emissions to the atmosphere. A 'relevant statutory provision' is The Alkali etc. Works Regulation Act 1906 which, as extended by Orders of 1966 and 1971, empowers the HSE, through its Alkali Inspectorate, to control a wide range of emissions from the workplace.

Local authorities, by reason of their powers under the Public Health and Clean Air Acts, also control atmospheric emissions. Generally speaking, the division of responsibility is such that local authorities concern themselves with emissions of smoke, dust and grit (as from combustion processes) and bring proceedings for statutory nuisances; whilst the Alkali Inspectorate of the HSE deal with emissions from industrial processes that are chemically complex or difficult to control.

The matter generally, however, entails the measurement of pollutant gases, vapours and suspended matter as an essential part of the system of control. In some instances, as for acid gases, prescribed limits are set. To comply with these implies presumptive evidence that the 'best practicable means' of emission control are in use.

6 The United States of America

A statement of the US Supreme Court in 1898 established the health of the worker as an objective of public policy. Legislation to control occupational health appeared first in the State of Massachusetts in 1905, and many other states, but by no means all, followed suit in the next few years. The content of the state laws varied considerably, as did their enforcement, which was generally tardy owing to the employment of untrained inspectors, often on a political basis.

During the first two decades of the twentieth century, much attention was attracted to problems of lead poisoning and several states adopted laws concerned specifically with this subject. Powers of enforcement were vested in the Department of Labor at the state level.

During the second decade, workmen's compensation laws were adopted by many states and in the 1930s occupational diseases became compensable under them. This led to a rapid increase in concern with occupational hygiene, which was to accelerate through Federal Government action during the years of the Second World War. Many industrial hygienists thereby obtained training and experience and after the war became engaged by state agencies to pursue their profession. Federal Government influence was however sustained through the requirements of the Walsh–Healy Act affecting government contract work. These included the first statutory reference to the Threshold Limit Values of the American Council of Governmental Industrial Hygienists (ACGIH) as a numerical expression of permitted atmospheric conditions in the workplace.

The legislative control of industrial health continued to be primarily at State level for another 20 years. It was therefore uneven and far from complete, both geograpically and with regard to its scope. Even more was this the case with industrial safety. Other important influences were at work however. Large insurance groups such as the National Fire Protection Association, the National Board of Fire Underwriters, the Chicago Underwriters and others, were very active in undertaking research and testing and in setting industrial safety standards that transcended state, and even national boundaries. Much safety research and information collecting was carried out by Government agencies, such as the Bureau of Mines and the Bureau of Standards, by academic and professional bodies and by industry itself. Thus the USA became a repository for information on the science and practice of industrial safety on a scale not exceeded elsewhere. By the end of the 1960s, the Federal Government was prepared to intervene comprehensively with national safety and health legislation. The outcome was The Coal Mine Health and Safety Act of 1969 and the Occupational Safety and Health Act of 1970. The former, *inter alia*, set, on an interim but nevertheless mandatory basis, specific numerical levels with respect to health (coal mine dust, oxygen deficiency and toxic gases) and safety (coal mine dust and flammable gas) and required the monitoring of the atmosphere of coal mines in these respects.

In 1977, the Federal Mine Safety and Health Act was passed by Congress, resulting in the transfer of the Mine Enforcement of Safety Administration from the US Department of the Interior to the US Department of Labor and the creation of the Mine Safety and Health Administration.

The Occupational Safety and Health Act is, of course, of wide significance in the American scene and is comparable in scope with the Health and Safety at Work Act in the United Kingdom. Its administration is the joint responsibility of the Department of Labor and the Department of Health and Human Services. The agencies concerned are respectively the Occupational Safety and Health Administration (OSHA) and the National Institute for Occupational Safety and Health (NIOSH).

The Department of Labor has the primary responsibility for policy and enforcement of the Act. The Department of Health, Education and Welfare is responsible for providing support through investigation and research, the publication of information and the supply of adequately trained manpower.

The initial purpose of the Federal Government is to achieve uniformly high standards in health and safety throughout American industry by centralised initiative. It is however the future purpose, as stated in the Act, to devolve powers of enforcement upon the individual states as and when they have developed legislation, policies and resources adequate for the new purposes. Federal agencies will however remain available to maintain standards and to deal with special situations.

The principal mechanism for establishing specific regulations is through 'consensus standards' agreed by regulatory agencies, industry and professional institutions. The Threshold Limit Values of the ACGIH are an example of such standards. It is the intention to develop 'criteria documents' for individual materials and hazards, leading to definitive standards. These will aim to define and quantify hazards as the result of the fullest possible investigation and to recommend protective procedures, including methods of measurement of, for example, levels of hazardous (toxic or flammable) materials in working atmospheres. An example is the criteria for a recommended standard on occupational exposure to carbon dioxide (HEW Publication No. (NIOSH) 76–194, August 1976), which includes detailed procedures for sampling and analysis of the gas in air, a problem specifically referred to in the Coal Mine Health and Safety Act.

An official publication of considerable importance and value is *The industrial environment – its evaluation and control* (NIOSH, 1973). This is a training textbook giving guidance to the regulations and official recommendations, and including much detail on the sampling and analysis of hazardous gases in general.

For the future, it is planned that, while minimum standards will be set at the level of the Federal Government, detailed methods of compliance will be settled within the State jurisdiction at the most local level that is practicable.

7 References

1. C. T. Coathupe, *Essay on the comparative effects of human respiration, and of the combustion of charcoal upon the atmosphere; and on the practical influence of atmospheric changes, produced by certain processes in manufactures, upon the health of those who are engaged in them.* Philp and Evans, Bristol (1838).
2. J. A. Wanklyn and W. J. Cooper. *Air analysis: a practical treatise on the examination of air, with an appendix on illuminating gas.* Kegan Paul, Trench, Trubner, London (1890).
3. J. S. Haldane. *Methods of air analysis.* Charles Griffin, London (1912).

4. F. W. Hardwick and L. T. O'Shea. Notes on the history of the safety-lamp. *Trans. Inst. Min. Engrs.*, **51**, 548–703 (1915–16).
5a. W. Forstmann. 'Die Verschiedenen Bauarten von Wetteranzeigern'. *Glückauf*, **60**, 1008–18 (1913).
b. H. Schultze-Rhonhof. Zusammenfassende Datstellung aller bisher zum Nachweis von Grubeugas in Bergwerken in Vorschlag gebrachten. *Glückauf*, **60**, 415–20 (1924).
c. Miners' Lamp Committee. *Memorandum No. 10.* HMSO, London (1924).
d. H. F. Coward and C. S. W. Grice. Principles of self-acting firedamp alarms. *Trans. Inst. Min. Engrs.*, **81**, 350–7 (1931).
e. L. C. Ilsley and A. B. Hooker. Evolution of methane-detecting devices for coal mines. *US Bur. Mines Inf. Cir. 6733,* 12 pp (1933).
f. F. J. Hartwell. Recent developments in methanometry. *Min. Fuel Power, Safety in Mines Research and Testing Branch Res. Rept. 9.* (1950).
g. E. J. Gleim. Portable methane-detecting appliances approved under United States Bureau of Mines Standard. *US Bur. Mines Rept. Invest. 5056* (1954).
6. W. R. Clanny. On the means of procuring a steady light in coal mines without the danger of explosion. *Phil Trans.*, 200 (1813).
7. H. Davy. On the firedamp of coal mines and on methods of lighting the mines so as to prevent its explosion. *Phil. Trans.*, 1 (1816).
8. J. J. Forbes and G. W. Grove. *USBM Miners' Circular 33* (1948).
9. G. Broomhead and S. R. Howarth. An investigation of the gas caps produced by methane in flame lamps. *SMRE Res. Rept. 281* (1972).
10. M. D. Monnier. Methanometre automatique en analyseur automatique du Grison. *Arch. Sci. Phys. Nat. Geneva 3rd ser.*, vol. 5, 331–9 (1881).
11. C. McLuckie. The detection of inflammable gases and vapours. *Trans. Inst. Min. Engrs.* **79**, 282–7 (1929–30).
12. J. Crossland. The Ringrose firedamp indicator. *Trans. Inst. Min. Engrs.* **76**, 6–16 (1928).
13. E. H. Liveing. On a new instrument for the detection and measurement of inflammable gas in mines. *Phil. Mag.* **9**, 126–31 (1880).

2 Measurement of flammable gases and vapours

J. G. Firth

1 Introduction

The reliable detection and measurement of concentrations of flammable gases and vapours in the working environment is one of the most effective methods for reducing the risk of fires and explosions. Many leakages of flammable gases or vapours occur under conditions where they are not obvious; for example, leakage from a flange in a gas pipe could result in large accumulations of flammable gas or vapour. In such situations, a continuously operated gas detection system can allow remedial action to be taken to minimise the accumulation of gas before the concentrations in the atmosphere around potential igniting sources reach a flammable level. In addition, when men must enter areas where flammable gas concentrations may exist to do work and hence provide igniting sources, then the only way of checking whether or not gas at flammable concentrations is present is by use of a flammable gas detector.

All flammable gases have a concentration in air below which they cannot be ignited called the lower explosive limit (LEL). This concentration is characteristic of an individual gas but the values for most flammable gases fall within the range 0.5% v/v to 13% v/v.

The spread of flame through a flammable gas/air mixture is governed by a number of factors which affect heat transfer, such as turbulence, as well as the basic chemistry of the oxidation reaction. Hence the LEL of a gas is not a fundamental physical constant but is affected by the way in which it is measured. For example, the lower explosive limit for hydrogen in air, when the flame is travelling upwards through the gas mixture, is 4% v/v, whilst the LEL for a mixture through which the flame is travelling downwards is 8% v/v. LELs are measured in standardised apparatus[1] in which the flame travels upwards through the gas/air mixture.

It is broadly true that a flammable gas/air mixture is safe as long as the concentration is below the LEL. Above this concentration, a hazardous condition exists where almost any igniting source, such as an electrical spark, will cause an explosion. Hence, most flammable gas detectors are designed to measure a concentration range up to the LEL of a gas so that in effect they measure the approach towards a hazardous condition at the measuring point.

The two requirements for inspection and monitoring have resulted in the development of three basic forms of instrument. These are portable spot-reading instruments, portable continuously operating monitors, and fixed continuously operating monitors. The first type of instrument is generally a small, self-contained device, which samples the atmosphere either by diffusion or by aspiration and gives an almost instantaneous indication of the concentration at the measuring point. Flame safety lamps would be included in this category but most modern devices display the concentration on a meter scale or on a digital read-out. These devices are normally capable of some hundreds of readings before they require any attention.

Fixed continuous monitors are installed on factory sites and monitor more or less continuously the concentration levels at strategic points within the site. Such points would be those where a leakage of gas is likely to occur, for example, over a flange in a pipe or, at points where possible igniting sources are present such as electrical equipment. Any change in flammable gas concentration to a level which is thought to indicate dangerous conditions in that particular situation will trigger an alarm signal or automatically apply remedial action, such as shutting off possible igniting sources or closing down part of the industrial process.

Portable continuous monitors combine the functions of the two types of devices described above. They are carried by operators on to their working site. They are used where the work process of the operator could cause leakage of flammable gases or vapours and/or provide a source of ignition. Since they can be read by the operator whilst he is working, they can be used by him to check that the work location is safe before he commences operation. They also provide a continuous check on whether or not flammable gases or vapours are entering the work location, and they generally have both a local alarm and a concentration read-out.

All gas-detecting instruments are composed of four basic units. These are shown in Fig. 2.1.

(1) The sensing unit, which detects the gas to be measured and produces some form of analogue signal. This sensing unit can be an electrocatalytic element, a spectroscopic unit consisting of gas cell radiation source and detector or, in the case of a flame safety lamp, the flame of the lamp.

(2) A central processing unit, which governs the power supply to the sensors and often processes the signal from the sensor for display to the operator.

(3) A display unit, which presents the signal in an appropriate form to the operator and/or takes appropriate action. This unit would include alarm functions, meter displays, automatic cut-off, etc.

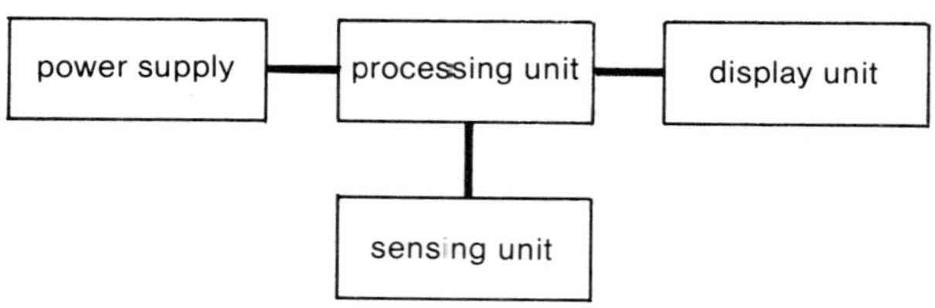

Fig. 2.1 Instrument components.

(4) A power supply for the three units. This is generally a battery in portable equipment or a mains supply in continuously operating fixed equipment. In the latter case, a floating battery is often included in case of mains failure or if the mains power is shut off when flammable gas is present.

The complexity and multiplicity of these units in a given gas detection system are obviously governed by the applications of that system. In a simple portable spot-reading instrument, a single sensing unit, possibly based on a catalytic sensor, will be used with a power supply consisting of a small battery, a display unit of a simple meter and a processing unit made up of a few resistors. However, in a system protecting a large industrial site, such as an oil rig, many hundreds of sensing systems will be operated, and a processing unit can be coupled to each sensing unit or one central processing unit may control all the sensing units. The display unit can be a display for each of the sensing units coupled with a multiple alarm system, with alarms situated at strategic points around the plant. The power unit will be coupled to the main electrical supply, with possibly floating batteries for each of the sensing units in the event of power supply failure.

It is obvious that for a detection system to work, the atmosphere being monitored must come into contact with the sensing unit. This can be achieved either by placing a sensing unit at the point to be measured or by transporting the atmosphere from that point to the unit, along a pipe. However, in the latter case, there must necessarily be a time delay between changes in concentrations at the measuring point being detected by the sensor. In some applications, gas concentrations will only build up slowly when a leak occurs and such time delays are then not important. However, in many applications, leakage of gas could result in very rapid increases in concentration and it is necessary to give as early a warning as possible. In these applications obviously the sensor unit should be mounted at the point of measurement. Such applications call for sensing units which are rugged and small and have led to the development of solid state sensors. However, where response time is not so important, more fragile systems such as spectrometers can be used and the gas may be piped to them from sampling points which are analysed in turn by the spectrometer.

Of the four basic units which constitute a gas detection system, by far the most important unit is the sensing unit. This is because the performance of modern electronic components which comprise most of the other three units is far in advance of the performance of the sensing units used in modern instruments. Therefore the performance of the system is as good or as bad as the performance of that part of the sensing unit which converts the gas concentration into some form of signal. Hence, the major part of this chapter is concerned with the characteristics of the sensing systems because by understanding these one can understand the advantages and limitations of the measuring system as a whole.

2 Sensing systems

2.1 Flame lamps

This was the first form of flammable gas detection system and is still widely used in

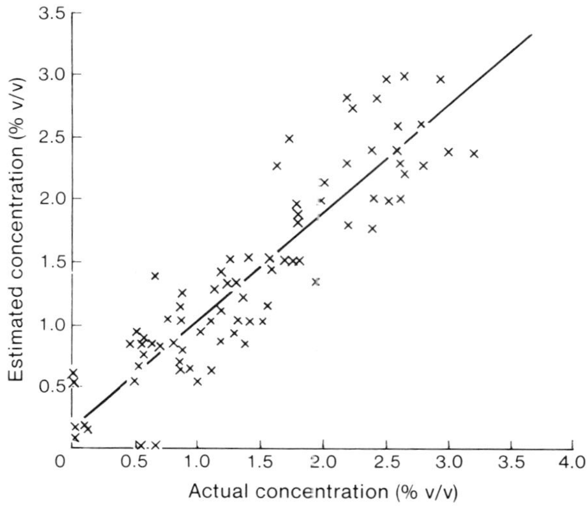

Fig. 2.2 Methane measurements by trained observers with flame safety lamp (ref. 3).

different parts of the world in mining operations. In these operations, firedamp, which is mainly methane, is released from the strata that are disturbed by the working. Although mainly used in coal mining, the flame lamp is also used in other locations where methane is the predominant component of the flammable gas to be measured, for example, gas from rotting organic material in sewers, and gas released from adjacent coal strata in civil engineering operations.

In order to measure methane concentrations, the flame is turned down to a minimum height until a small yellow 'eye' is left in the paraffin flame. The methane then in the atmosphere under test burns as a small blue triangular-shaped flame on top of the paraffin flame. The shape and height of this flame is a function of the concentration of methane in the atmosphere.[2]

The interpretation of these flame caps is a subjective assessment on the part of the operator and will be affected by the frequency with which he observes flame caps produced from calibrated atmospheres, i.e. by training, and by the colour perception of the operator in the blue part of the spectrum. Factors such as age and fatigue, all affect the latter. As with many systems that involve individual operator skill, considerable folklore has grown up about the reliability of flame lamp measurements and the accuracy with which certain individuals can interpret the flame caps. However, at best, a flame lamp is an inaccurate way of measuring methane concentrations and considerable variation exists in the readings obtained by trained operators[3] (Fig. 2.2).

Nevertheless, the flame lamp has a number of advantages. It is cheap, simple to operate and perhaps most important, it is easy to check that it is operating properly, i.e. that the flame is either in or out. It is, however, being rapidly replaced on a world-wide basis by instruments which do not require skilled subjective assessment of concentrations.

Various modifications have been made to the basic flame lamp system over the years to reduce the subjectiveness in assessing concentrations. These have generally involved using some form of temperature sensor to measure the temperature increase produced by the burning of methane in the flame. In most cases a thermocouple is positioned above the flame to do this.[4] The positioning of the thermocouple in the hot gas stream from the flame is critical; hence, as would be expected, such modifications are markedly orientation sensitive. Although not very satisfactory in portable systems such modifications have had some degree of success in fixed continuous monitoring systems, where the lamp can be rigidly fixed in place. The most notable of these[5] utilised a butane flame which was controlled by a valve system which ensured that the gas flow to the flame, and hence the flame size, was independent of environmental variables such as temperature, pressure, etc. The temperature rise produced by methane is relatively small, e.g. 1% methane produces a 5% change in temperature. The fluctuations in the flame itself cause sufficient background noise on this signal from the gas to limit the accuracy to about 20%. Maintenance of this type of device is also complex and most of these modifications of the basic flame lamp system have been superseded by devices based on catalytic sensors.

2.2 Catalytic detectors

To some extent this type of sensor is an extension of the flame lamp, in that flammable gas is oxidised to carbon dioxide and water and the heat emitted in the process is measured. However, by using a solid catalyst, the temperature at which the oxidation takes place is markedly reduced and the process can take place on a small solid-state electrically operated sensor. The catalyst is incorporated into an electrically heated sensor containing a temperature-sensitive device and the flammable gas is oxidised continuously on the catalyst. The heat evolved raises the temperature of the catalyst and this temperature rise is measured by the heat sensor.

In most devices, the temperature sensor is a platinum resistance thermometer, which is wound in the form of a coil and is also used as the electrical heater for the device. The most suitable catalysts for promoting the oxidation of flammable gases are metals in Group 8 of the Periodic Table and of these, platinum and palladium are by far the most widely used. Hence, the oldest form of these sensors is simply a small coil of platinum wire. The gas oxidises on the surface of the platinum and raises the temperature of the coil and the resulting resistance change in the coil is measured by using it as one arm of a Wheatstone bridge and measuring the out-of-balance voltage produced across the bridge (Fig. 2.3).

A typical sensor would be a 15-turn coil of 75 μm diameter wire with a length of 5 mm and a diameter of 0.5 mm. This sensor is still widely used in many portable spot-reading instruments. However, it is not suitable for use in continuously operating devices because platinum is a relatively poor catalyst for the oxidation of flammable gases, especially methane. In order to oxidise methane, these sensors must be operated at temperatures greater than 800 °C to obtain sufficient sensitivity. At this temperature, platinum is evaporating from the coil. This evaporation is increased markedly in the presence of flammable gas.[6] Hence, the

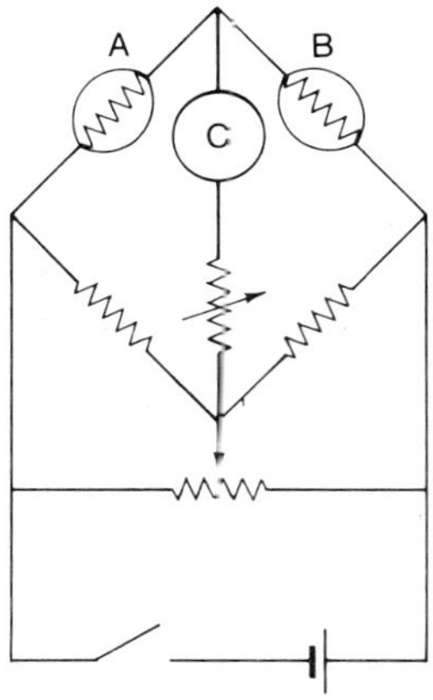

Fig. 2.3 Wheatstone bridge network used with catalytic elements. A, catalytic element; B, compensating element.

cross-section of the coil is continually changing, resulting in a continual change of electrical resistance and a continual change in the zero of the instrument. This process results in early burn-out of the device.

Attempts to overcome this deficiency of the simple platinum coil have centred on the use of more active catalysts for the oxidation of flammable gases and the design of transducers which enable these catalysts to be used in conjunction with a platinum coil resistance thermometer. Palladium and its oxides are much more active catalysts than platinum, especially for methane, oxidising it at temperatures 300 °C below that necessary for platinum. A range of transducers[7,8] has resulted which are all basically similar in concept to that shown in Fig. 2.4. The platinum coil acting as a heater and as a temperature sensor is either wound round a support or embedded in a porous ceramic bead and the catalyst is impregnated on to the outside of either the support or the bead. The encapsulation or support of the coil makes it very resistant to changes in geometry produced by mechanical shock and the spherical symmetry of the device makes it insensitive to change in orientation.[9] Such devices are used in the vast majority of flammable gas detection instruments.

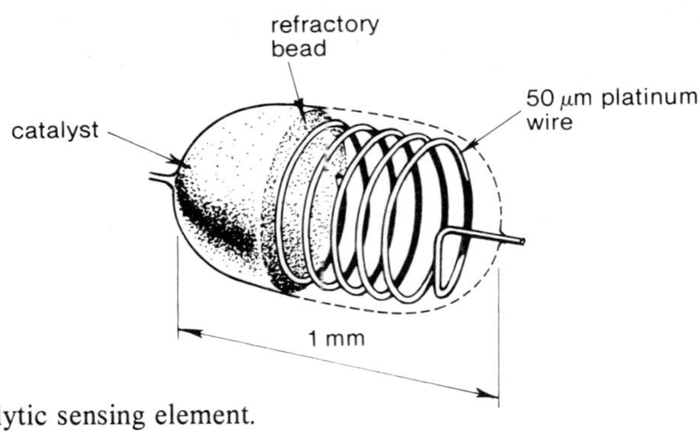

Fig. 2.4 Catalytic sensing element.

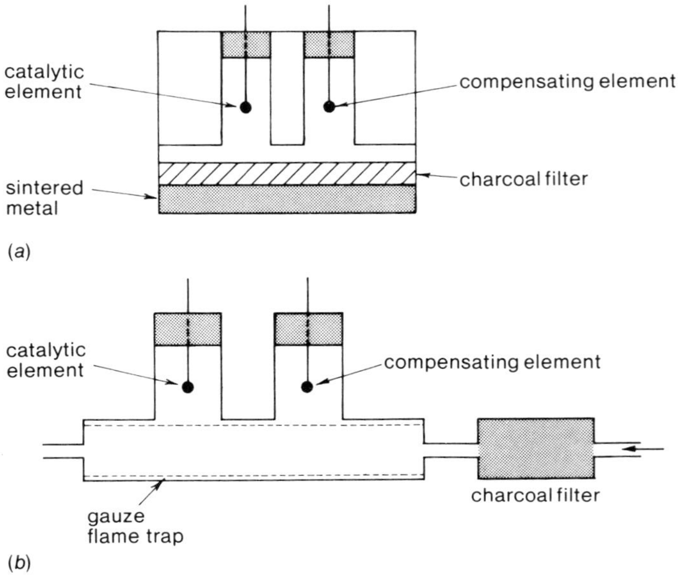

Fig. 2.5 Gas sampling systems for catalytic elements. (*a*) Diffusion sampling system; (*b*) aspirated system.

In addition to the transducer itself, the other main component of the sensing unit is the gas flow system, which enables the gas to reach the transducer. Such systems are of relatively simple design (Fig. 2.5). The element is contained in a blind cavity or similar system where the transport of gas to the element is controlled by diffusion. In many instruments, the transport of gas from the outside atmosphere into the cavity itself is controlled by diffusion (Fig. 2.5*a*). In some portable instruments, gas flows past the end of the cavity under the influence of an electrical pump or a simple aspirator (Fig. 2.5*b*), and then diffuses from the gas stream up the cavity to the sensing element. The main feature of the design of the cavity is that the walls and ends should be sufficiently far away from the sensing element to ensure that they do not interfere with the convection currents around the hot elements. In this way, heat losses from the element are not affected by changes in orientation of the instrument, and hence the zero of the instrument and the sensitivity are unaffected by orientation changes. This is generally achieved if the distance from the element to the cavity walls is about five times the radius of the element.

In operation of the instrument, the electrical power to the bridge will heat the sensing elements to their operating temperature of about 500°C. The bridge is balanced in air and the zero voltage across the bridge is the zero signal of the instrument. On exposure to flammable gas, the heat of oxidation of the gas on the element raises its temperature, which in turn increases the resistance of the thermometer inside the element, causing an out-of-balance voltage across the bridge; this out-of-balance voltage is calibrated in terms of gas concentration. At low concentrations of flammable gas ($<40\%$ LEL), the out-of-balance voltage on the bridge is a linear function of gas concentration.[10]

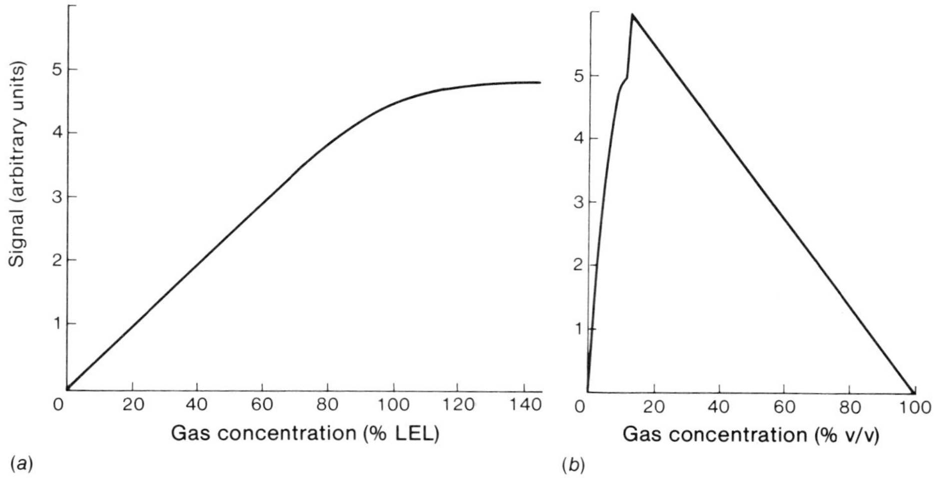

Fig. 2.6 Signals from catalytic elements.

However, as the flammable gas concentration is increased, two factors begin to affect the linearity of the signal. As more and more gas oxidises on the surface of the catalyst, a greater fraction of the catalyst area is used in the reaction. The signal will eventually level out as the catalyst surface area becomes saturated.[11] This is the normal effect observed in heterogeneous reactions controlled by Langmuir–Hinshelwood or Eley–Rideal kinetics.[12] Thus, the signal concentration curve will have the form shown in Fig. 2.6(*a*).

Where flammable gas concentrations are being measured in air, then increases in concentrations of the flammable component result in a decrease in the concentration of oxygen in the atmosphere. Eventually, insufficient oxygen will be present in the atmosphere to ensure complete oxidation of the flammable gas. Thus, at higher concentrations, the signal from the sensor will decrease as the flammable gas concentration increases, and have the general form shown in Fig. 2.6(*b*). At 100% flammable gas concentrations, the signal from the bridge will be zero.[11] In practice, this will occur only when characteristics of the elements governing heat losses are perfectly matched. In general, differences in size, colour, etc., which affect heat losses, result in a positive or negative signal in 100% flammable gas on account of the difference between the thermal conductivities of the gas and air.

In general, most instruments are used to measure concentrations up to the LEL of the gas of interest. The ambiguous signals which can occur in very high concentrations on account of lack of oxygen are only important with flammable gases that are permanent gases. However, since this includes such important substances as hydrogen (town gas) and methane (natural gas), many attempts have been made to design measuring systems without this basic deficiency. Most of these systems utilise the difference in thermal conductivity between the flammable gas and air. If a catalytically inactive element, heated from an almost constant power supply in air, is progressively exposed to increasing concentrations of flammable gas in air, then, if

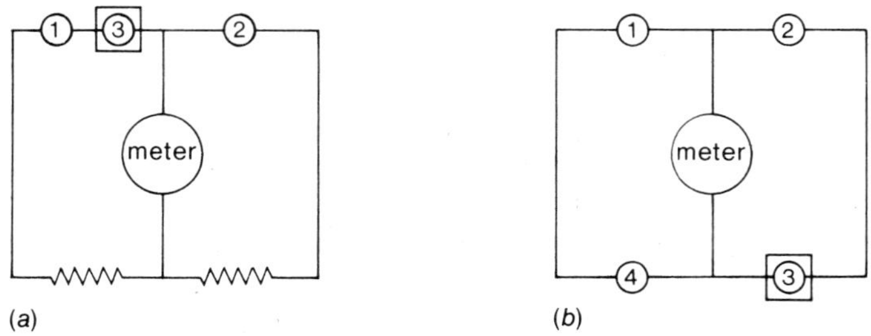

Fig. 2.7 Non-ambiguous catalytic/thermal conductivity bridge systems. (*a*) 1, catalytic element; 2, compensating element; 3, isolated compensating element. (*b*) 1, catalytic element; 2, compensating element; 3, isolated compensating element; 4, compensating element.

the thermal conductivity of the flammable gas is greater than that of air, the temperature of the element will decrease monotonically up to 100% flammable gas. Where the element is contained in a Wheatstone bridge network, then out-of-balance signals will be obtained. This effect is utilised[13] in the systems described in Fig. 2.7. In each case, element 3 is isolated and is not exposed to the atmosphere under test. Elements 1 are catalytically active, elements 2 are inactive compensating elements, and in Fig. 2.7(*b*) element 4 is a further compensating element. The effect of this in both cases is to produce an increasing thermal conductivity signal in high concentrations of gas, which is superimposed on a decreasing catalytic signal (Fig. 2.8). If the instrument has a full scale reading of 100% LEL corresponding to point A, concentrations of gas greater than 100% LEL will always produce a signal which is off scale.

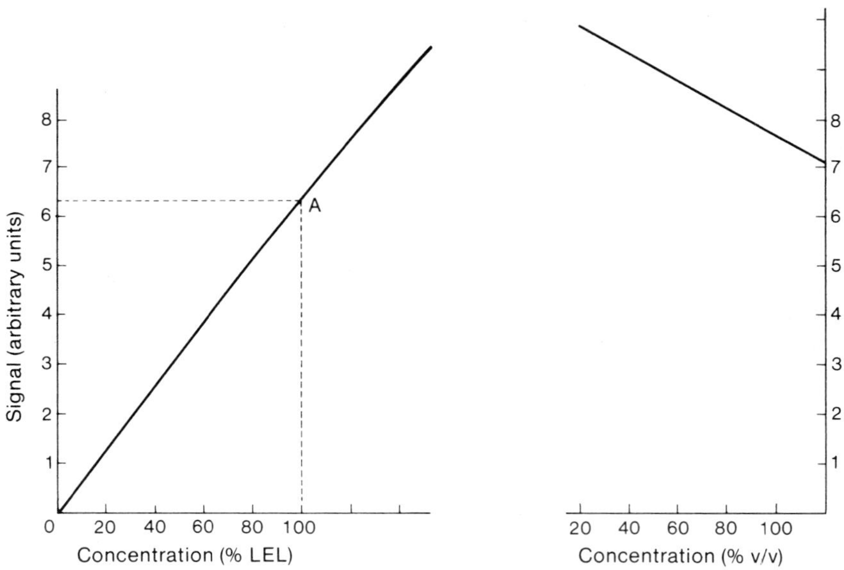

Fig. 2.8 Signal from catalytic/thermal conductivity bridge system.

Later developments of this use of thermal conductivity systems to overcome ambiguity problems have utilised separate catalytic bridges and thermal conductivity bridges.[14] In these systems, the thermal conductivity bridge can be used to measure over the scale 0–100% by volume of the flammable gas and the signals from the thermal conductivity bridge and the catalytic bridge can be combined electronically to produce an unambiguous signal measuring in the range up to the LEL of the gas.

It must be emphasised that the systems described above can only be used to measure flammable gases consisting of basically one component, e.g. natural gas. They cannot be used where a mixture of flammable gases is present, especially if the composition of the mixture can vary and the thermal conductivities of the components are markedly different. The problems of ambiguity are really only of importance where a portable instrument is used to probe for pockets of flammable gas. It can be assumed that fixed monitoring instruments will normally be operating in concentrations well below the LEL and rises in concentrations towards the LEL will result in alarms and subsequent action.

The size of the signal from a catalytic transducer in a given concentration of flammable gas will be governed by the rate of oxidation of that gas on the catalyst and the heat of oxidation. The rate of oxidation of any gas on a catalyst is governed by the surface area of the catalyst and the kinetics of the particular reaction. There will obviously be variations in the surface area between transducers and wide differences in the kinetics of the reaction for different gases at a given catalyst temperature. Thus, in principle, the signals from different flammable gases at the same concentration will be different. This could limit the use of the instrument to one type of flammable gas and make it useless for measuring flammable gas mixtures of varying compositions.

This defect is overcome if the catalytic sensor is operated in conditions where the rate of reaction of the gas is not governed by the surface catalytic reaction but by ensuring that the rate at which the gas reaches the surface of the catalyst is governed by diffusion.[15] This can be achieved by limiting the size of the cavities containing the elements so that diffusion up the cavity controls the rate of oxidation of gas on the sensor. Alternatively, it can be done by operating the sensor at such a high temperature that a depletion layer is produced immediately above the surface of the catalyst and diffusion of flammable gas through this depletion layer is rate controlling. In practical instruments, it is probable that a combination of these two effects is used. Under these conditions, it can be shown that the signal produced across a Wheatstone bridge is given by Equation 2.1.[15]

$$V = K \cdot \Delta H \cdot D_{12}[F] \tag{2.1}$$

where V is the signal across Wheatstone bridge (mV);
ΔH is the enthalpy of oxidation of flammable gas ($\mathrm{J\,mol^{-1}}$);
D_{12} is the coefficient of diffusion of flammable gas in air ($\mathrm{cm^{-1}\,s^{-1}}$);
$[F]$ is the concentration of flammable gas;
K is a constant.

Table 2.1 Calculated (Equation 1) and experimentally observed signals for different gases

Gas	*Ratio signal on gas to signal in calibration gas*	
	Experimental value	*Calculated value*
Pentane[a]	1.00	1.00
Acetone	1.00	1.18
Benzene	1.10	1.02
Butane	1.00	1.15
Butadiene	1.54	1.30
Carbon disulphide	0.47	0.41
Carbon monoxide	1.75	1.75
Diethyl ether	1.10	1.05
Ethyl alcohol	1.30	1.30
Ethylene	1.92	1.64
Heptane	0.90	0.90
Hexane	1.00	0.94
Hydrogen	1.75	1.79
Methyl alcohol	1.54	1.67
Methyl ethyl ketone	0.91	0.96
Octane	0.80	0.88
Propylene oxide	1.18	1.07
Toluene	1.00	1.00
Xylene	0.80	0.92

[a] Calibration gas.

This equation applies to the approximately linear portions of the signal curve shown in Fig. 2.6, that is up to and beyond the LEL. Good agreement has been found between the signals predicted by Equation 2.1 and the experimentally determined values (Table 2.1).

Under diffusion controlled conditions, all the flammable gas reaching the surface of the catalyst is oxidised. Hence, when a mixture of flammable gases is being measured, the signal from the sensor will be the sum of the signals which would have been produced by the oxidation of each component of the mixture separately (Equation 2.2).

$$V = \sum_i K_i \cdot \Delta H_i D_i [F_i] \tag{2.2}$$

where V is the signal across Wheatstone bridge (mV);
K_i is the constant for gas 'i' of mixture;
ΔH_i is the enthalpy of oxidation of flammable gas 'i' in air (J mol^{-1});
$[F_i]$ is the concentration of flammable gas 'i';
D_i is the coefficient of diffusion of gas 'i' in air (cm^{-1} s^{-1}).

Over a given catalyst, different gases will oxidise at different rates at the same catalyst temperature. As the temperature of the catalyst is increased, then for each gas the rate of reaction will increase, until diffusion controlled conditions set in. This

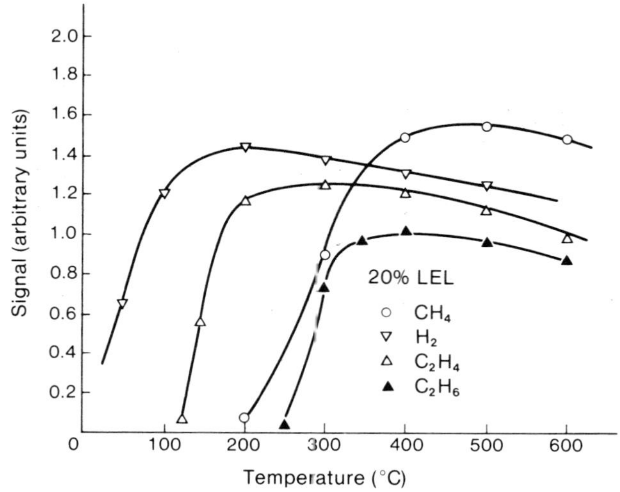

Fig. 2.9 Effect of element temperature on catalytic bridge signals.

effect is shown in Fig. 2.9. Hence, if the temperature of the catalyst is sufficiently high to ensure that the most difficult to oxidise gas is being oxidised under diffusion controlled conditions, then the other components of the mixture will also be oxidised under the same conditions. In general, methane is the flammable gas that is most difficult to oxidise. Any instrument, whose sensor is operated so that the oxidation of methane is diffusion controlled, will also oxidise the vast majority of other flammable gases. The validity of Equation 2.2 for a sensor operating in this manner for a mixture of gases with widely ranging individual or explosive limit values is shown in Table 2.2.

It is perhaps a statement of the obvious that a flammable gas detector is intended to measure the hazard from combustible gases. However, flammable gases are a hazard only at or above the LEL. Since the LEL for different gases differs widely, e.g. 5% v/v for methane and 1.1% v/v for hexane, it is an advantage if the concentration of the gas shown by the instrument is expressed as a fraction of the LEL of the gas and is independent of the gas being measured. Catalytic sensors approach this desirable situation because for most gases, the product of the heat of oxidation and the LEL is approximately constant.[16] This relationship is illustrated for ninety-three flammable gases in Fig. 2.10. Hence, substitution into Equation 2.1 shows that the output from the catalytic sensor can be expressed in terms of the gas concentration as a percentage of the LEL of the gas (Equation 2.3).

$$V = K_2 D_{12} [F]/[\mathrm{LEL}] \qquad (2.3)$$

where V is the signal across Wheatstone bridge (mV);
K_2 is a constant (mV.cm.s);
$[F]$ is the concentration of flammable gas;
[LEL] is the concentration of flammable gas at LEL;
D_{12} is the coefficient of diffusion of gas in air ($cm^{-1}\ s^{-1}$).

In fact Equation 2.3 shows that, if the diffusion coefficient for each gas was constant, then the signal for almost any flammable gas would be the same at a given fraction of its LEL. If this were so, then the instrument could be calibrated with any flammable gas and the calibration would hold for most of the gases. The function $D_{12} \cdot \Delta H$ has been calculated as a function of the LEL for a large number of gases and is shown in Fig. 2.11. The departure from linearity when compared with Fig. 2.10 shows the effects of the diffusion coefficient of the gas on the signal from the instrument. If an instrument were calibrated with pentane, then the signal from any gas would lie on the line, if it was independent of the nature of the gas. Departures from this line are a measure of how far an instrument would be in error if it were calibrated with pentane and then used to measure other gases. In practice, calibration factors are provided with an instrument which has been calibrated in a particular gas, to enable it to be used in other gases. Table 2.3 shows the correction factors which must be used for a range of flammable gases when an instrument is calibrated with pentane. Many gases (8–20) can be estimated reasonably well (±20%) with an instrument calibrated against pentane, without using a correction factor. But a number of common gases (1–7) will give high readings, whilst others (21–26) will give low readings. More accurate readings will be obtained with the former gases, if the instrument is calibrated with hydrogen, and with the latter gases, if it is calibrated with hexane.

Table 2.2 Calculated (Equation 2.2) and experimentally observed signals for gas mixtures

Gas	*Conc (% v/v)*	*LEL (% v/v)*	*LEL mixture (% v/v)*	*Observed signal (mV)*	*Calculated signal (mV)*
Methane	0.5	5.3	4.6	19.0	20.0
Hydrogen	0.5	4.0	—	—	—
Methane	0.33	5.3	—	—	—
Hydrogen	0.33	4.0	3.4	21.1	19.4
Propane	—	—	—	—	—
Carbon monoxide	3.0	12.5	9.8	43.2	42.8
Methanol	1.9	7.3	—	—	—
Carbon monoxide	1.5	12.5	—	—	—
Methanol	1.9	7.3	8.3	43.4	43.3
Methane	0.5	5.3	—	—	—
Methane	0.5	5.3	4.2	18.8	19.4
Ethane	0.5	3.0	—	—	—
Methane	0.33	5.3	—	—	—
Ethane	0.33	3.0	3.1	23.2	21.5
Propane	0.33	2.2	—	—	—
Carbon monoxide	1.5	12.5	5.1	44.1	37.1
Hexane	0.5	1.1	—	—	—
Carbon monoxide	1.5	12.5	—	—	—
Hexane	0.5	1.1	3.7	43.1	37.6
Methane	0.5	5.3	—	—	—

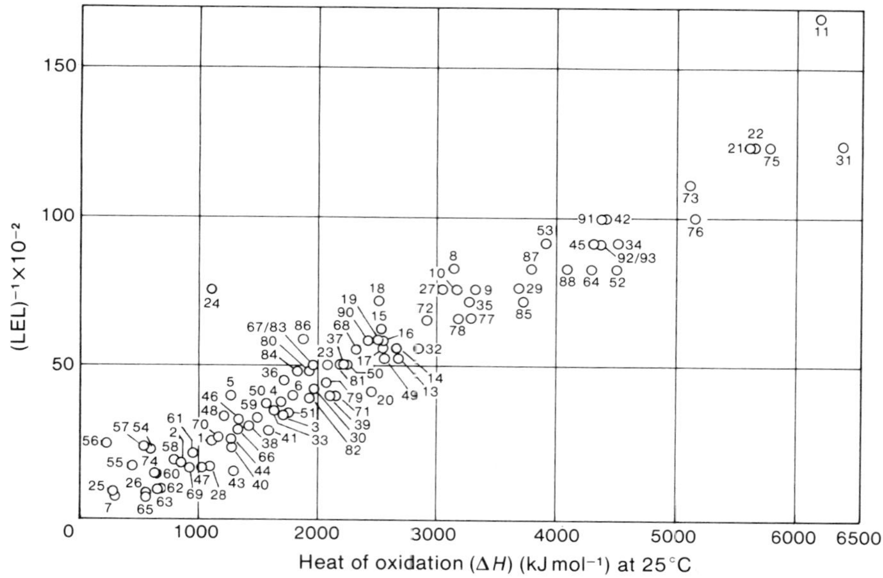

Fig. 2.10 Relationship between heat of oxidation and lower explosive limit (LEL). 1. acetaldehyde, 2. acetic acid, 3. acetic anhydride, 4. acetone, 5. acetylene, 6. allyl alcohol, 7. ammonia, 8. amyl alcohol, 9. aniline, 10. benzene, 11. biphenyl, 12. 1,3-butadiene, 13. butane, 14. isobutane, 15. butene-1, 16. *cis*-butene-2, 17. *trans*-butene-2, 18. butyl alcohol, 19. *iso*-butyl alcohol, 20. *tert*-butyl alcohol, 21. butyl benzene, 22. *iso*-butyl benzene, 23. butyric acid, 24. carbon disulphide, 25. carbon monoxide, 26. carbon oxysulphide, 27. chlorobenzene, 28. cyanogen, 29. cyclohexane, 30. cyclopropane, 31. decane, 32. diethylamine, 33. dimethylamine, 34. 2,3-dimethylpentane, 35. 2,2-dimethylpropane, 36. dimethyl sulphide, 37. 1,4-dioxane, 38. ethane, 39. ethyl-acetate, 40. ethyl alcohol, 41. ethylamine, 42. ethylbenzene, 43. ethyl bromide, 44. ethyl chloride, 45. ethylcyclopentane, 46. ethylene, 47. ethylene dichloride, 48. ethylene oxide, 49. diethyl ether, 50. ethyl formate, 51. ethyl mercaptan, 52. heptane, 53. hexane, 54. hydrazine, 55. hydrogen cyanide, 56. hydrogen, 57. hydrogen sulphide, 58. methane, 59. methyl acetate, 60. methyl alcohol, 61. methylamine, 62. methyl bromide, 63. methyl chloride, 64. methylcyclohexane, 65. methylene dichloride, 66. dimethyl ether, 67. methyl ethyl ether, 68. methyl ethyl ketone, 69. methyl formate, 70. methyl mercaptan, 71. methyl propionate, 72. methyl-propyl ketone, 73. naphthalene, 74. nitromethane, 75. nonane, 76. octane, 77. pentane, 78. 2-methyl butane, 79. propane, 80. propyl alcohol, 81. propylamine, 83, propylene, 84. propylene oxide, 85. *iso*-propyl ether, 86. propyne, 87. toluene, 88. triethylamine, 89. trimethylamine, 90. vinyl ethyl ether, 91. *o*-xylene, 92. *m*-xylene, 93. *p*-xylene.

The situation is more complicated when mixtures of flammable gases are to be measured. In principle, an instrument which is to be used for measuring the hazard from a gas mixture should be calibrated with that mixture. If, however, the instrument is calibrated with pentane, then high readings will be obtained from mixtures where each component will give a high reading and low readings will be obtained from a mixture where each component will give a low reading. This is illustrated in Table 2.4.

However, if a gas which gives a low reading, for example hexane, is mixed with one which will give a high reading, for example carbon monoxide, then the signal

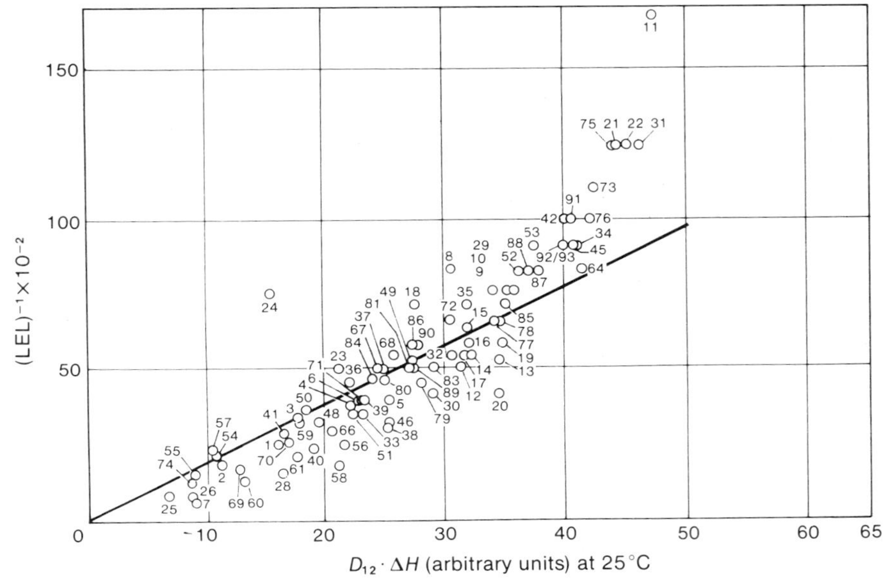

Fig. 2.11 Relationship between lower explosive limit and $D_{12}\,\Delta H$. Numbers as in Fig. 2.10.

obtained from the detector will depend markedly on the relative concentrations of each gas. In such cases, if the ratio of the components is fixed, then the instrument can be calibrated with the mixture. However, if the ratio of components can vary, then the signals from a catalytic explosive gas detector should be interpreted with caution.

The effects of temperature and pressure changes on the signals from catalytic sensors, have been discussed.[17] It has been shown that, where an instrument has been calibrated at a temperature T_0 and is then used in an atmosphere at a higher temperature T_1, a correction factor can be used with the reading at the higher temperature to obtain a true measure of the gas concentration expressed as a fraction of its LEL. The correction factor is shown in Equation 2.4.

$$X = T_1/T_0\,[1-6.6\times10^{-4}(T_1-T_0)]^{-1} \qquad (2.4)$$

where X is the correction factor;
T_0 is the calibration temperature (°C);
T_1 is the measuring temperature (°C).

The factor takes into account that the LEL of the gas is temperature dependent. It is useful where an instrument can only be calibrated at room temperature and is then to be used in situations where the temperature is markedly higher or lower. Where the signal from an instrument is controlled by diffusion, changes in absolute pressure have little effect on the accuracy. The diffusion coefficient of the gas is inversely proportional to the absolute pressure[18] and the number of molecules entering the diffusion zone of the instrument is directly proportional to the absolute pressure.

Hence the signal from the sensor is almost independent of the pressure of the gas mixture being measured. Since the LEL of the gas expressed as percentage by volume is almost independent of the absolute pressure,[1] the reading given by the instrument is almost independent of the absolute pressure. In practice, a small degree of pressure dependence is observed, the signal being directly proportional to the absolute pressure, with a proportionality constant of approximately 2.5×10^{-2} when the pressure is expressed in torr and the signal in percentage LEL. Thus, for an increase in pressure of 100 torr, the signal from the gas concentration of 40% LEL at atmospheric pressure will rise to 41% LEL.

The major problem associated with the use of instruments employing catalytic sensors is the possibility that the sensor may encounter vapours which will either decompose on the catalyst to produce solid products which cover the catalyst surface, or will be so strongly absorbed by the catalyst that they take up reaction sites on the catalyst which would be used by the flammable gas or oxygen in their reaction. When this occurs, the signal in a given concentration of gas will be reduced and the instrument will be failing to register danger. The types of vapours which can

Table 2.3 Correction factors using different calibration gases

No.	*Gas*	*LEL (%v/v)*	*Correction factor for different calibration gases*		
			Pentane	*Hydrogen*	*Hexane*
1	Methane	5.3	0.51	—	—
2	Methanol	7.3	0.54	0.89	—
3	Hydrogen	4.0	0.61	1.00	—
4	Carbon monoxide	12.5	0.62	1.01	—
5	Ethanol	4.3	0.63	1.04	—
6	Ethylene	3.1	0.66	1.07	—
7	Ethane	3.0	0.68	1.12	—
8	Butane	1.8	0.79	—	—
9	Acetylene	2.5	0.81	—	—
10	Propane	2.2	0.84	—	—
11	Acetone	2.6	0.90	—	—
12	Ethyl acetate	2.5	0.90	—	—
13	*cis*-Butene-2	1.7	0.95	—	—
14	Pentane	1.5	1.0	—	—
15	Diethyl ether	1.9	1.0	—	—
16	Methyl ethyl ketone	1.8	1.12	—	—
17	Benzene	1.3	1.13	—	—
18	Cyclohexane	1.3	1.13	—	—
19	Toluene	1.2	1.15	—	—
20	Xylene	1.1	1.19	—	—
21	Heptane	1.2	1.20	—	0.96
22	Hexane	1.1	1.25	—	1.00
23	Octane	1.0	1.24	—	0.99
24	*n*-Butyl alcohol	1.4	1.35	—	1.07
25	*n*-Amyl alcohol	1.2	1.41	—	1.13
26	Nonane	0.8	1.47	—	1.17

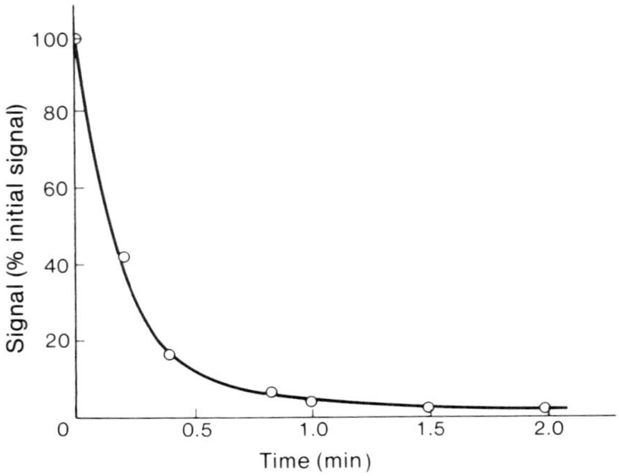

Fig. 2.12 Effect of silicone on catalyst sensitivity. Poison concentration: 40 p.p.m. (hexamethyl disiloxane).

decompose on the catalyst to produce solid products are those containing silicon, lead and phosphorus. Silicones are widely used in industry in potting compounds, greases and polishes. The decomposition of silicones to form silicon oxides on catalyst surfaces is cumulative and a few parts per million of the vapour can rapidly render the catalyst inactive (Fig. 2.12). Silicone vapours are the most potent agents for

Table 2.4 Effect of calibration gas on reading in gas mixtures

Gas mixture	*Composition (% v/v)*	*LEL mixture (% v/v)*	*Signal at 50% LEL for different calibration gas*	
			Pentane	*Hydrogen*
Methane	50	4.6	87	52
Hydrogen	50	—	—	—
Methane	33	—	—	—
Hydrogen	33	3.4	72	43
Propane	33	—	—	—
Carbon monoxide	61	9.8	86	50
Methanol	29	—	—	—
Carbon monoxide	39	—	—	—
Methanol	49	8.3	93	55
Methane	12	—	—	—
Methane	50	4.2	79	47
Ethane	50	—	—	—
Methane	33	—	—	—
Ethane	33	3.1	74	44
Propane	33	—	—	—
Carbon monoxide	60	—	—	—
Hexane	20	3.7	82	49
Methane	20	—	—	—

reducing catalyst activity but lead tetraethyl in gasoline and organic phosphorus compounds used as corrosion inhibitors are also effective. The effect is permanent so that, when the element is removed from the atmosphere containing the poison, it does not recover its activity.

Gases which contain halogen atoms, such as carbon tetrachloride, ethylene bromide, etc., also reduce catalytic activity, since, when the gas is adsorbed on the surface of the catalyst, it is decomposed and the resultant halogen atoms are strongly adsorbed on the surface. Again, a few parts per million of halogen-containing gas can reduce the catalyst activity to almost zero. However, the halogen is desorbed from the catalyst so that removal of the element from the containing gas vapour leads to recovery of the catalytic activity. The only way to check whether a sensor has been poisoned in this way or not is to expose it to flammable gas.

A catalyst surface may contain many different sites which will catalyse oxidation of flammable gases, even though only one material, e.g. palladium, is used as a catalyst. Some flammable gases may use all types of site and some may use one or a few types of site in their oxidation reaction. Similarly, poisons may be adsorbed on one or more types of site. Thus, a molecule which may poison the oxidation of one flammable gas, may not poison or only partially poison, the oxidation of a second gas.[19] This is illustrated in Fig. 2.13. Hence, it is important when checking to see if a sensor is poisoned, that it is exposed to the gas which the instrument is to measure and not to some other flammable gas which may still oxidise on the sensor.

Ways of overcoming the effect of catalyst poisons depend on preventing the poison molecules from reaching the catalyst surface. The earliest devices simply used a charcoal filter contained either as a cartridge in the gas stream of an aspirated system or as a filter layer in a diffusion system. The charcoal strongly adsorbs most catalyst poisons and prevents them from reaching the catalyst. Such devices can be very effective; for example, a layer of charcoal of thickness 2 mm, exposed continuously to a silicone atmosphere containing 5 p.p.m. (a very high concentration) of silicone, will last for one month before the silicone starts to reach the

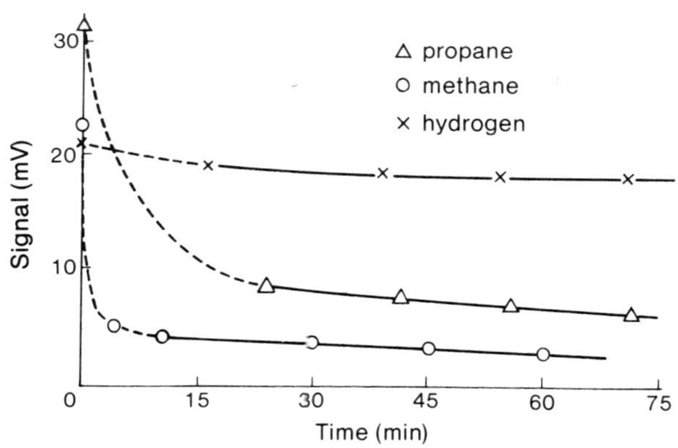

Fig. 2.13 Effect of silicone poisoning on sensitivity of different gases to oxidation. Catalyst: Pd/ThO_2. Temperature: 600 °C. Poison: 40×10^{-6} v/v hexamethyl disiloxane.

catalyst.[11] However, these systems have the disadvantage of preventing the vapours of substances which are liquid or solid at room temperature from reaching the catalyst. Thus, they cannot be used where flammable gases, with a molecular weight greater than that of propane, are being measured.

Alternative filter systems have been developed, where the hot sensor is enveloped in a plug of fibrous silica or asbestos.[20] The inner parts of this envelope are heated by a sensor and the poison molecule decomposes on this heated plug before reaching the catalyst surface. Such systems are claimed to be at least an order of magnitude more resistant to poisons than an unprotected sensor.

Later attempts to overcome the defect have concentrated on producing truly poison-resistant catalysts. Catalytic systems using ion-exchange zeolites have also been developed.[21] Zeolites are alumino-silicates with a three-dimensional internal pore and cavity systems, in which the diameter of the pores is of the order of a few Angstroms. Since many poison molecules, especially silicones, tend to be large, they cannot enter the internal pores system of the zeolite. Smaller flammable gas molecules can. Zeolites are ion-exchange materials and palladium, platinum etc. ions, which are catalytically active, can be placed in the internal pore structure of the zeolite. Such catalysts have markedly increased resistance to poisons (Table 2.5). Current work is aimed at producing zeolite structures which have increased thermal stability at the operating temperature of the sensors (600 °C) to prevent loss of sensitivity when the device is operating in air. This loss is brought about by slow thermal decomposition of the zeolite structure.

A more recent development in poison resistant sensors involves the formation of a small catalytic bead which is encapsulated in a highly porous bead of alumina. The porous bead allows the passage of flammable gas through to the catalytic sensor; the poison molecules are decomposed within the alumina. Such beads involve the use of catalysts with proven stability and have markedly improved resistance to catalyst poisons. Highly porous alumina/palladia catalysts are also used.[22]

2.3 Semiconductor sensors

Recent years have seen the introduction of solid state sensors for flammable gases

Table 2.5

Catalyst poison	*Zeolite pore size (nm)*	*Catalytic ion position*	*Ratio of activity for methane oxidation after exposure to initial activity*
Hexamethyl disiloxane	1.0	outside pores	0.1
Hexamethyl disiloxane	1.0	inside pores	0.8
Carbon tetrachloride	1.0	outside pores	0.05
Carbon tetrachloride	1.0	inside pores	0.0
Carbon tetrachloride	0.4	outside pores	0.1
Carbon tetrachloride	0.4	inside pores	1.0
Chlorine	0.4	inside pores	0.0

which are based on metal oxide semiconductors. As with catalytic devices, they rely on the adsorption of a gas on to a heated oxide surface. The adsorption and/or subsequent reaction of a flammable gas on the surface of the oxide produces an electrical conduction change in the metal oxide itself on account of the electronic processes involved in the reaction on its surface. These conductivity changes are related to the amount of gas adsorbed on the surface of the oxide and hence to its concentration in the surrounding atmosphere. These effects have been known for many years[23] and have often been used to study adsorption processes on solids and the surface species produced in catalytic reactions.[24] Adsorption of a gas on to the surface of a solid involves the formation of chemical bonds. This necessarily involves a disturbance of the electron distribution in the solid surface. The simplest case is the adsorption of a gas molecule to produce a charged ion on the surface of the solid. The abstraction or donation of an electron from the solid to the gas molecule to produce the ion will produce a conductivity change within the solid.

Oxides of the transition metals and heavy metals, such as tin, are the most widely used semiconductors in this type of sensor. These oxides are semiconductors on account of the ability of the metal to exist in different oxidation states. All of the oxides are non-stoichiometric and contain a slight excess (n-type semiconductor) or deficiency (p-type semiconductor) of metal ions. In ionic solids, the discreet and isolated electron levels of the individual ions are broadened into electron energy bands which extend throughout the solid (Fig. 2.14). In a stoichiometric oxide the band containing electrons in the highest energy levels will be completely filled (valence). The next higher energy band (conduction) will be completely empty. Net movement of charges in the valence band is not possible and there are no charge carriers in the conduction band. Electrons cannot be thermally activated into the conduction band from the valence band because the energy gap is too large (generally of the order of 3 eV). Hence, such a material would be an insulator. Most of the transition metal oxides have an excess (n-type) or deficiency (p-type) of metal ions in their lattice.

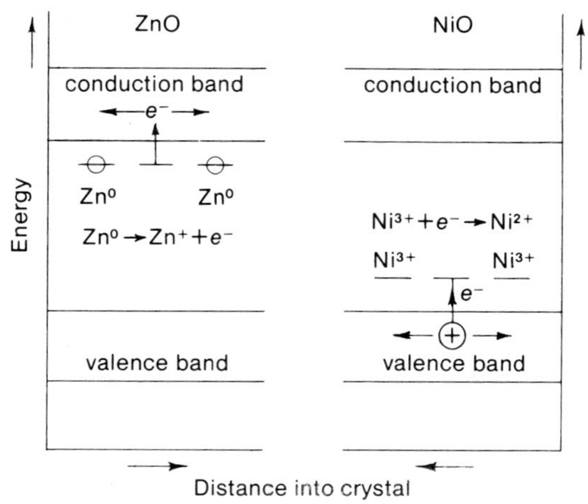

Fig. 2.14 Band structure of oxide semiconductors.

It is necessary in any metal oxide to preserve electrostatic neutrality; hence, in a material with an excess of metal ions present, some of the metal ions must be present in a lower oxidation state than those of the metal ions of the lattice, e.g. Zn^0 in Zn^{2+}. Such ions have localised energy levels just below that of the conduction band of the oxide and these localised electrons can easily be thermally excited into the conduction band. Once there, they are free to move throughout the oxide and impart electrical conductivity to it. In the process, of course, the Zn^0 ions are converted to Zn^{2+} ions. Similarly, in materials with a deficiency of metal ions, in order to preserve electrostatic neutrality, some of the metal ions are present as ions with a higher oxidation state than those of the lattice metal ions; for example, Ni^{3+} in NiO. These ions have localised energy levels which are just above the top of the valence band of the oxide. Electrons can be thermally excited from the valence band into these localised levels, leaving positive holes in the valence band which are free to move through this band and impart electrical conductivity to the oxide.

The simplest mechanism by which adsorption of the gas in an ionic form can produce a change in the conductivity of an oxide is by changing the concentration of those metal ions in the oxide lattice which have a charge different from that of the main metal ion in the lattice. Thus, adsorption of neutral oxygen molecules on to the surface of an oxide to form two oxygen ions results in the transfer of four electrons from the solid to the adsorbed gas, where they are localised. On an n-type semiconductor oxide such as zinc oxide, these four electrons will have come from two Zn^0 ions, converting them in the process to Zn^{2+} ions. Thus, the number of ions contributing to the conductivity of the metal oxide will be reduced and adsorption of oxygen on an n-type semiconductor will cause a decrease in its conductivity.

Adsorption of oxygen on to a p-type semiconductor, such as nickel oxide, will again result in the localisation of four electrons on the two oxygen ions produced on the surface. These four electrons will arise from the conversion of four Ni^{2+} to four Ni^{3+} ions. This increase in the concentration of ions contributing to the electrical conductivity of the oxide results in an increase in conductivity when oxygen is adsorbed on to a p-type metal oxide. These effects are summarised in Fig. 2.14.

A combustible gas can be adsorbed on to the surface of a transition metal oxide to produce neutral fragments of the original molecule or to produce positively charged ions (e.g. CH_3^+). The species produced will depend on the nature and the temperature of the oxide. If the sensor is used in air, then oxygen ions will be present on its surface on account of the adsorption of oxygen from the gas phase. These oxygen ions will react with the adsorbed combustible gas molecule/ions, to form combustion products, such as carbon dioxide and water, which will be desorbed from the surface. The overall effect is the removal of adsorbed oxygen ions from the surface and hence the electrons which were associated with these ions will be donated back to the oxide lattice to become associated with the localised energy levels and hence re-available for the conduction process. Hence, adsorption of a combustible gas will result in an increase in electrical conductivity of an n-type semiconductor and a decrease in the conductivity of a p-type semiconductor.

In an atmosphere containing both oxygen and combustible gas, oxygen will re-adsorb on the surface so that a steady state concentration of oxide ions will be present on the surface of the oxide and the conductivity will be determined by the gas phase concentration of the gases. However, in general, oxygen is strongly adsorbed on metal oxides so that above a minimum concentration the surface is saturated and marked changes in oxygen concentration produce little change in the conductivity. Combustible gases, however, are generally weakly adsorbed so that the number of adsorbed species increases with increasing gas concentration. Since the concentration of charge carriers in semiconductors is relatively small, large relative changes in their concentration are produced by adsorption of combustible gases. This means that relatively large changes in conductivity can be produced by small concentrations of gas. For example, 500 p.p.m. of methane will change the conductivity of stannic oxide at 250 °C by a factor of 10.

From the description above, it is obvious that any gas which can be adsorbed on to a metal oxide and cause an electron redistribution in the solid surface will produce a change in conductivity. Non-combustible gases, such as sulphur dioxide, nitrogen dioxide and water, will also produce large conductivity changes. Thus, both combustible and non-combustible gases can produce signals in devices utilising metal oxide semiconductors. This lack of specificity towards combustible gases is a major limitation in the use of such devices for combustible gas detection.

It will also be apparent that the signal which will be produced by a gas is totally dependent on the surface reactions of that gas and hence will depend on the particular gas being measured, the oxide being used and its temperature. Thus, there is no relationship between the signal produced by a metal oxide and the lower explosive limit of the gas. This means that semiconductors cannot be used in simple explosimeters which need to be used for measuring different gases in the same way that catalytic sensors can.

The rate at which the equilibrium conductivity in a particular gas mixture is reached will depend on the rate of adsorption and desorption of the different gases on the oxide surface. Similarly, the rate at which the signal returns to zero when the combustible gas is removed will depend on the rates of desorption from the surface. For some gases these can be slow processes. Thus, on stannic oxide at 200 °C, the adsorption of methane and sulphur dioxide occurs in a few seconds but although the re-adsorption of oxygen is rapid, so that reversal of the signal on removal of methane takes place in a few seconds, the desorption of sulphur dioxide is slow and may take up to 10 minutes. Hence, it is possible for devices based on metal oxides, to have a slow response to the gas and a slow return after exposure to the gas. Since all these processes are activated, an increase in temperature of the oxide will decrease both the response time and the recovery time of a device utilising them.

The reaction and adsorption of a gas on the surface of an oxide will be affected by temperature changes. Moreover, the conductivity of the oxide itself will show an exponential dependence on temperature. Hence, the signal from a given gas will have a complex temperature relationship which will be characteristic of an individual gas (Fig. 2.15), so that changes in temperature of the oxide can produce a change in the relative signal from different gases. In general, metal oxide semiconductors are operated at much lower temperatures than catalytic devices, e.g. 200 °C. This gives them good long-term stability of both zero and sensitivity.

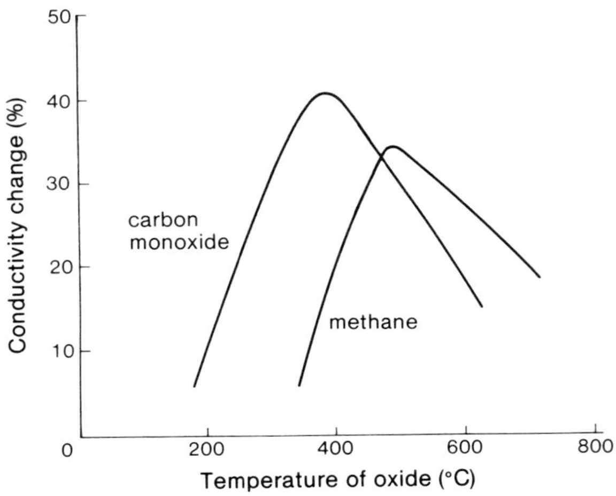

Fig. 2.15 Effect of temperature of oxide on conductivity change.

Moreover, most of the molecules which will poison a catalytic sensor will not decompose at these temperatures on metal oxides. Thus, their signal in a flammable gas is not affected by such molecules.

Hence, semiconductor sensors are devices with good long-term stability and large response to relatively small concentrations of flammable gas. However, they do not give a signal which is related to the explosiveness of the gas and they can give signals when non-combustible gases are present. This latter fault is a failure to safety. Thus, it would seem that they can mainly be used where low concentrations of flammable gas, i.e. tens of parts per million, would indicate fault conditions and normally the atmosphere in the area would be fresh air.

It will be obvious from the above description that the main requirements for the construction of a semiconductor gas sensor are an ability to heat a metal oxide to an appropriate temperature and a means of measuring its electrical conductivity. Since they will be devices that are heated, the considerations of spherical symmetry and

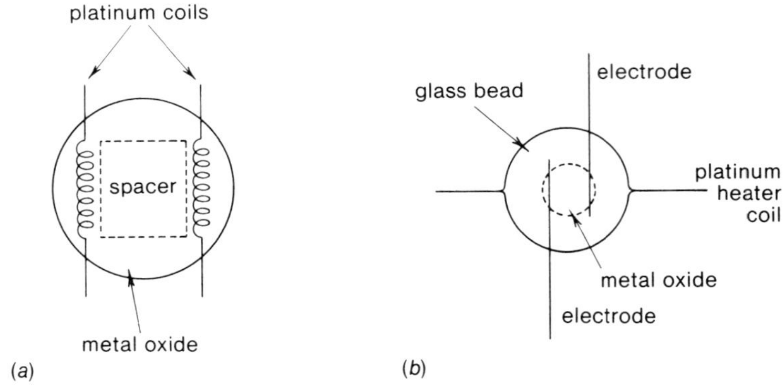

Fig. 2.16 Semiconductor sensors.

other factors which affect heat losses are the same as those described for catalytic sensors. Similarly, power requirements must be minimised so that the related circuitry can be made intrinsically safe. One of the earliest forms of sensor is that described by Taguchi,[25] in which two parallel platinum wire coils (length 5 mm, diameter 0.5 mm) are embedded in a bead formed from stannic oxide and silica (Fig. 2.16*a*). One of the coils is used to heat the oxide electrically to its operating temperature and both coils are used as electrodes. Another form of device[26] uses the oxide as a thin deposit on the surface of the small glass bead (1 mm diameter) between two wire electrodes sealed into the surface of the bead, which contains a heater coil (Fig. 2.16 *b*). Later devices use oxide single crystals.[27]

2.4 Infra-red detectors

Most flammable gas molecules absorb radiation in the infra-red region of the electromagnetic spectrum. Absorption of the radiation in this energy range will cause changes in the vibration of the bonds between the atoms forming the molecule. This excitation is quantised and is characteristic of the individual chemical bonds within the molecule. Hence, an absorption spectrum will be produced which is characteristic of the chemical bonds within the molecule (Fig. 2.17). Such spectra are 'finger-prints' of individual molecules but molecules possessing similar chemical bonds within them, e.g. C–H, will have spectra with similar absorption bands.

If infra-red radiation is passed through a sample of gas then part of the radiation will be absorbed and the intensity of the radiation emerging from the gas sample will be reduced. The transmitted intensity is related to the incident intensity and the concentration of gas in the sample by the Beer–Lambert law (Equation 2.5).

$$I_\nu = I_0 \exp(-\varepsilon c d) \tag{2.5}$$

where I_ν is the intensity of transmitted radiation;
I_0 is the intensity of incident radiation;
ε is the molar extinction coefficient ($m^3\,mm^{-1}\,mol^{-1}$);
d is the length of gas path (mm);
c is the concentration of gas in cell ($mol\,m^{-3}$).

The basis of an infra-red gas analyser then is an infra-red radiation source, a cell in which the gas to be tested can be enclosed and a detector placed at the opposite end of the cell to the source.

Two basic types of system are used. A dispersive infra-red analyser is one in which the radiation from the source is dispersed through a prism or grating and a selected band of the spectrum coinciding with the absorption band of the gas to be measured is passed through the gas cell. Often a second band of radiation which does not coincide with an absorption band of a molecule is also passed through the gas cell as a reference. The ratio of the intensities of the transmitted radiations is then a measure of the gas concentration.

Non-dispersive infra-red analysers (NDIR) are those in which dispersal of the radiation into its different wavelengths is not used. Either all the radiation from the

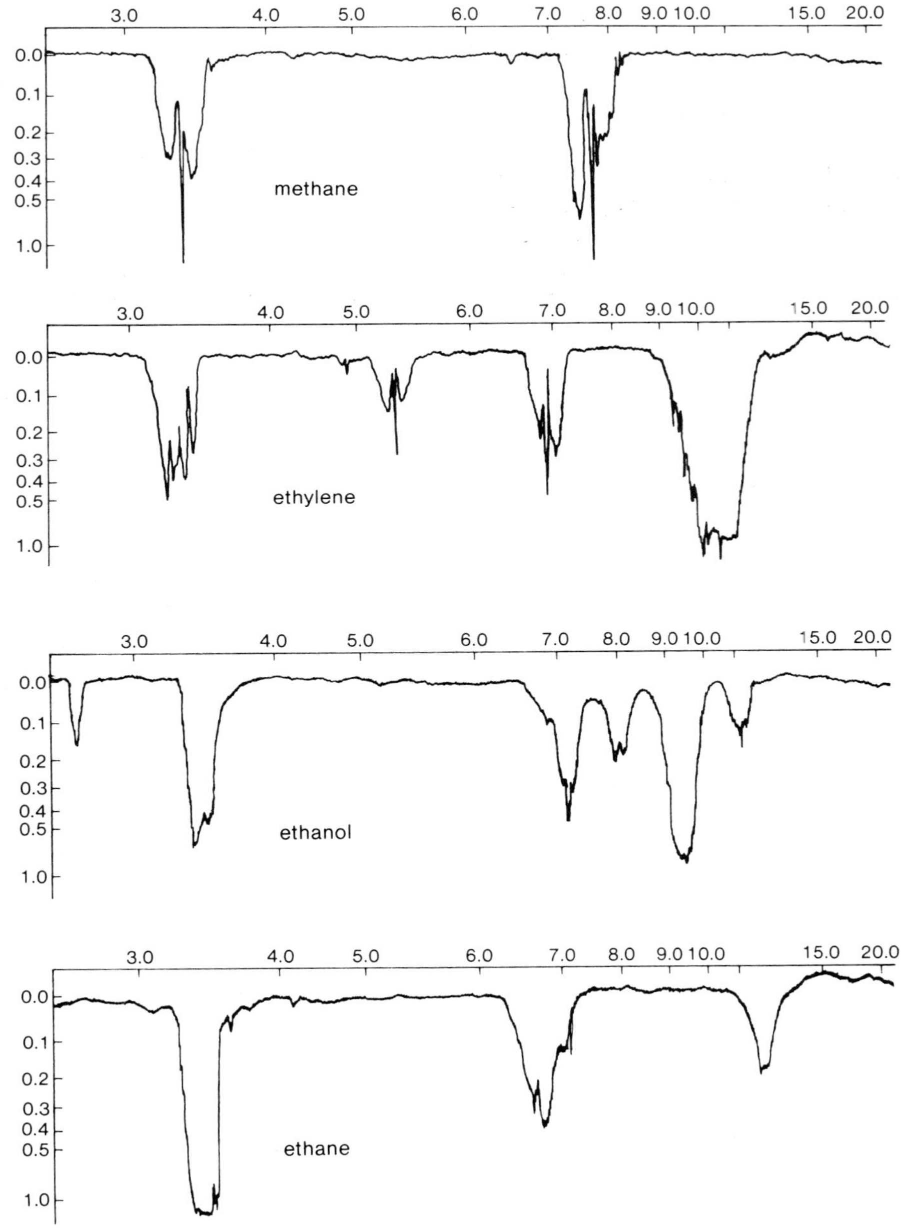

Fig. 2.17 Infra-red spectra.

source is passed through the gas or a narrow band of radiation produced using an infra-red filter is passed through the gas. In the former system, selectivity towards a gas is produced by the detector, whilst in the latter system it is produced by the filter.

In NDIR systems, even if all the absorption bands of a molecule are used, the

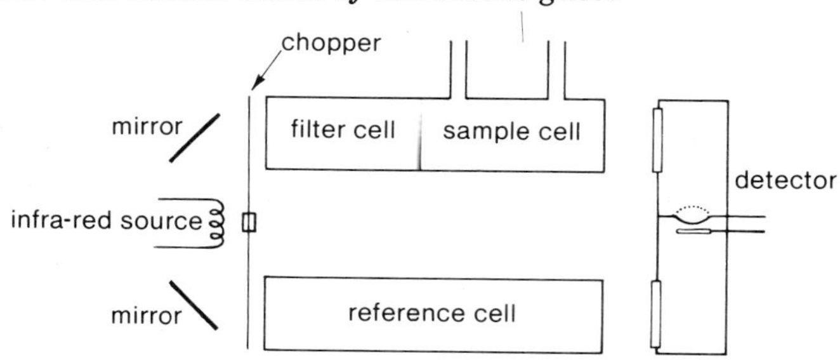

Fig. 2.18 Non-dispersive infra-red spectrometer.

changes in signal produced by the gas relative to the changes in signals produced by a drift in the emitter and detector, etc., are relatively small. Hence, reference signals which are subject to the same sort of drift, are used and the gas concentration is measured as a difference or a ratio of these signals. Reference signals are often obtained by passing radiation through a comparison cell into the same detector (Fig. 2.18).

The most widely used type of selective detector is based on the work of Veingerov[28] and Luft.[29] These detectors are gas-filled cells in which the heating effect of the infra-red radiation falling on the cells causes the gas to expand and the resultant expansion is measured. Selectivity is achieved because the gas in the cell will only absorb infra-red radiation in its own infra-red absorption bands and hence is basically only sensitive to changes in concentration of the same gas in a measuring cell. The principle of the Luft cell is shown in Fig. 2.19. The expansion of the gas in the cell displaces a diaphragm which is part of a condenser and the change in capacity of the condenser is measured electrically. The cell is divided generally into two portions, separated by the diaphragm. Transmitted radiation from the measuring cell falls on one portion of the detector cell and transmitted radiation from the reference cell falls on the other. When the intensities of these radiations are equal, the gas in both compartments expands to the same extent and there is no net displacement of the diaphragm. Changes in transmitted intensity from the measuring cell caused by the presence of gas there produce a movement of the diaphragm.

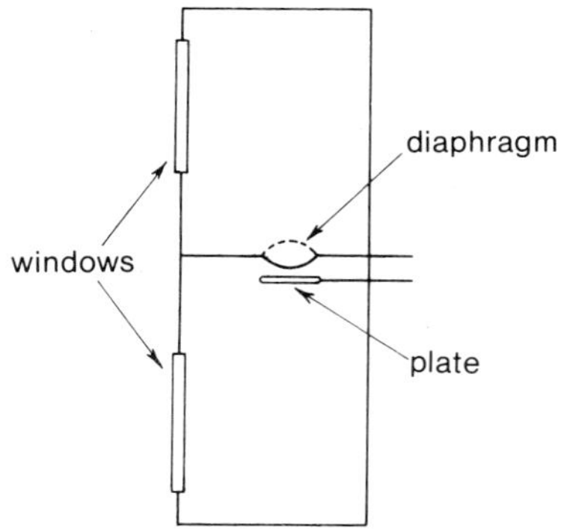

Fig. 2.19 Luft-type detector cell.

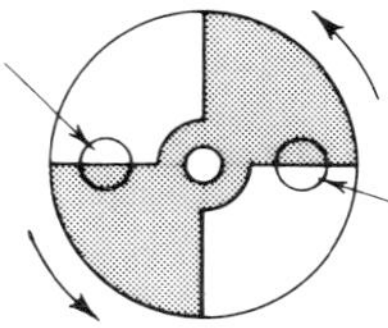

Fig. 2.20 Chopper. With balanced beams the areas of sample and reference cell covered are equal.

If the radiation falling on the cell were steady, the diaphragm would just move to an equilibrium position. Hence, a rotating mask (chopper) is placed in the radiation path (Fig. 2.20). This segmented disc rotates in a plane perpendicular to the radiation path so that the detector cell is subjected to regular pulses of radiation. This causes alternate expansion and contraction of the gas in the cell. When the radiation intensity falling on both sides of the cell is different, this causes the diaphragm condenser system to act as a microphone, producing a pulsating signal with a frequency twice that of the frequency of rotation of the chopper and a pulse height which is a function of the concentration of gas in the measuring cells. Various modifications to this basic system have been made to improve signal: noise ratio and hence sensitivity.

The NDIR type of system outlined above has a high degree of selectivity towards a particular gas on account of the almost perfect matching of the selectivity of the detector with the particular gas to be measured. Where the atmosphere under examination may contain gases with absorption peaks overlapping those of the gas of interest, the selectivity of the system as a whole can be further increased by putting cells containing these gases in the radiation path so that radiation in these overlapping bands is removed.

Most of the infra-red gas cells are constructed of stainless steel, brass or gold plated metals, with maximum reflectance within the cells. The cells may be embedded in metal blocks to increase thermal stability. Cell windows are generally made of calcium fluoride or a similar infra-red transparent material. The lengths of the cells are chosen to fit the concentration range in the gas to be measured. It is generally desirable to try to limit the absorption of the radiation to a range where it is almost linearly related to the concentration of the gas. In practice this is 5–20% absorption. The longer the optical path length within the cell, the lower the concentration of gas which can be measured. For flammable gases where concentrations in the low percentage range are to be measured, a path length of up to a few centimetres is sufficient.

Systems in which specificity is obtained, which use a filter system to isolate a radiation band which coincides with an absorption band of gas to be measured, have been gaining increasing attention. The development of interference filters[30] (Fig. 2.21) has enabled almost any narrow band of infra-red radiation to be selected. Indeed, continuously variable filters have been developed which enable the wavelength of the incident radiation to be selected at will within a particular instrument. Such filters use the constructive and destructive interference of

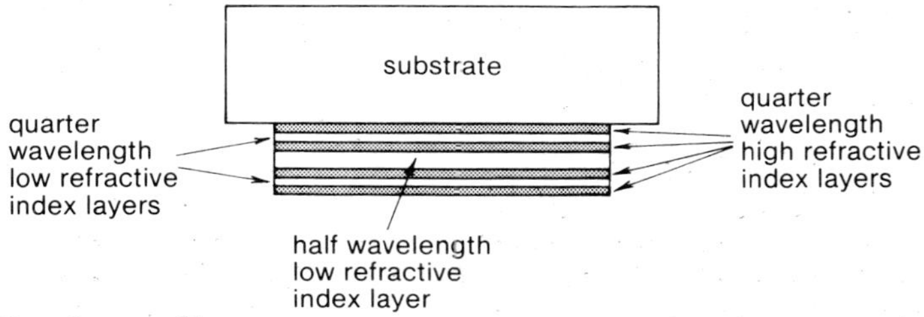

Fig. 2.21 Interference filter.

electromagnetic waves produced when they pass through a system of quarter wavelength thick layers of alternate high and low refractive index material and are used immediately before a non-selective detector.

Alternative narrow band sources of infra-red radiation are various laser systems. Of particular interest for flammable gases is the helium/neon laser which has an emission line at 3.39 μm. This coincides with a stretching frequency of the C–H bond which is present in most flammable gases. However, the width of the laser line is smaller than the absorption peaks in the fine structure of the C–H stretching band so that the absorption coefficient for the line varies from gas to gas (Table 2.6).

Table 2.6 Absorption coefficients for laser lines

Gas	*Laser line (μm)*		*Absorption coefficient at 1 atm (cm^{-1} atm^{-1})*
Methane	Ne	3.3913	15.0
	I	3.431	<0.5
Ethane	Ne	3.3913	4.4
	I	3.431	3.8
Propane	Ne	3.3913	8.9
	I	3.431	4.7
Butane	Ne	3.3913	13.1
	I	3.431	6.3
Pentane	Ne	3.3913	15.4
	I	3.431	9.0
Hexane	Ne	3.3913	17.9
	I	3.431	11.5
Acetaldehyde	Ne	3.3913	0.44
	I	3.431	0.39
	Xe	3.5070	1.6
	Xe	3.6849	2.1
Ethylene	CO_2	10.231	5.0
	CO_2	10.592	36.0
	CO_2	10.650	4.5
Methanol	CO_2	9.655	17.0

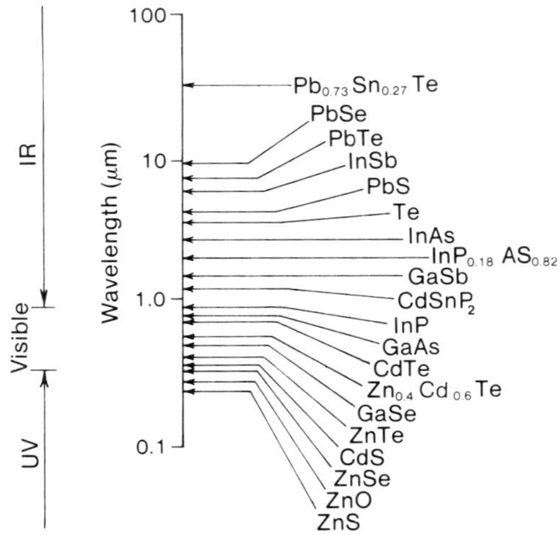

Fig. 2.22 Solid state laser systems – wavelength of laser emission.

Solid state laser systems have also been developed using cooled junction diode systems.[31] The wavelengths emitted by different systems are shown in Fig. 2.22. At the present time solid state lasers which will operate continuously at room temperature have not been developed but the eventual production of such devices will have a marked effect on gas detection.

Infra-red systems using interference filters or monochromatic radiation sources do not require selective detectors to produce specificity towards a gas and rugged solid state detectors can be used. Simple thermocouple detectors are available but most systems use photoconductive semiconductor detectors. In these devices the absorption of radiation excites charge carriers into or from impurity levels or produces a free-electron–free-hole pair. The effectiveness of such detectors is described in terms of the Detectivity,[30,32] D^* (dee-star), which is a measure of the sensitivity of the detector at a given wavelength. The detectivity for a given semiconductor material varies with wavelength, generally passing through a maximum (Fig. 2.23). Obviously, a detector is used whose maximum lies as close as possible to the absorption band of the gas which is being used.

The main advantage of an infra-red gas detection system is that the signal from the instrument is a function only of the concentration of the gas to be measured and is monotonic over the range 0–100% by volume. It is less subject to interference by other gases than the methods previously described and indeed can be highly specific for the particular gas to be measured. However, drifts in the system caused by deposition of dirt on the windows or reduction in output of the emitter all mean that for a given concentration of gas the signal is decreased so that the system tends to fail to register danger. In addition, an infra-red system is obviously much more complex and expensive than one based on catalytic sensors. Hence, it requires greater maintenance and has a higher initial cost. For these reasons, infra-red systems tend to be used only in situations where catalytic systems cannot be used. These would include situations where catalyst poisons are very likely to be present and where the

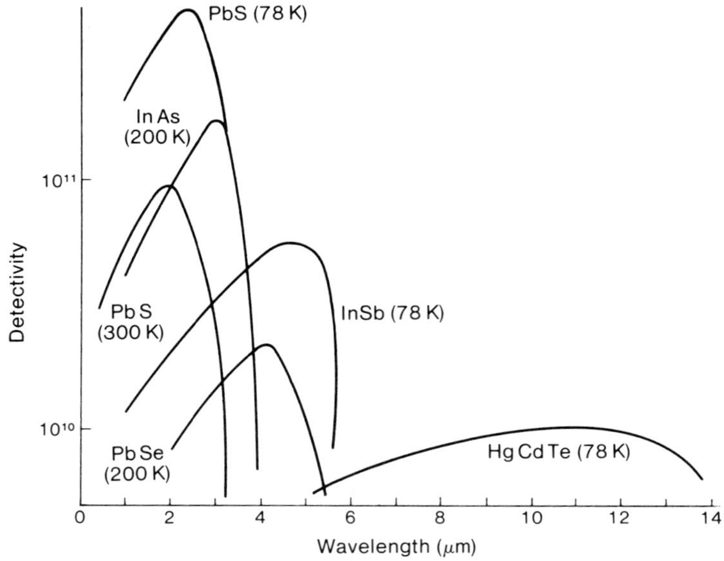

Fig. 2.23 Solid state infra-red detector materials (ref. 32 by permission American Chemical Society).

flammable gas is normally present in an inert atmosphere in which the oxygen content is insufficient for the catalytic sensor to operate. The high specificity of infra-red systems also limits their use in general flammable gas detection. Even if the cell of a Luft type detector is filled with a mixture of gases, that infra-red instrument can only be used for the gases in that cell. The output from an infra-red instrument can only be expressed in terms of the concentration of the gas or gas mixture for which the instrument has been designed. It will give a signal for other gases whose absorption bands overlap those for which the instrument is calibrated. The output from the instrument cannot, however, be interpreted in terms of the concentration of these gases, unless their composition is precisely known. Hence, the use of infra-red instruments should be limited to applications where the gas or mixture of gases which are to be detected is known and remains constant.

3 Standards of instruments

A number of countries have already set[33] or are considering setting standards for flammable gas detectors. These basically set out minimum accuracy and constructional safety requirements. Most industrial countries have, however, had similar standards for instruments in use in coal mining[34] for many years.

A standard for any instrument is determined by the job that the instrument has to do and the conditions in which it has to do it. Since these are basically similar throughout the world, most national standards will probably be similar to each other.

Flammable gas detectors do not need to be highly accurate (see p. 62). Most standards require an accuracy of about ±10% of the indicated reading or ±2% of the LEL, whichever is greater. However, the accuracy of the instrument has two components. The first is the inherent accuracy of the readings under standard conditions and the second is the effects on accuracy produced by changes in environmental variables. Some standards assess the combined effects of these variables statistically and set an overall requirement for accuracy such that 80% of the readings obtained by these shall be within the required accuracy limits. Other standards allow a maximum error in the reading for any range of each variable. The overall error then allowed ought to be smaller than the sum of the allowed errors for each variable. For example, a maximum error of 5% of the reading could be allowed for the range of each temperature, pressure, humidity, airspeed, etc. but the overall error should still be less than 10% of the actual reading.

A particular type of instrument may be used in a wide variety of applications so that the range of each environmental variable is made as wide as is reasonably practical. However, it would not be reasonable to cover extremes of operating conditions, e.g. arctic oil wells may require additional testing to be made on top of that for the certification process. The ranges of each environmental variable covered in standards are shown in Table 2.7.

Since flammable gas detectors are to be used in situations where flammable gas is present they must be safe to use in such situations. This means that they must meet the electrical safety requirements set out by individual countries. These requirements are detailed and varied for different countries, but basically the instrument must either be intrinsically safe or flameproof. The former normally requires that, if two simultaneous faults occur in the electrical circuitry of the instrument, then there shall be no possibility that it could ignite flammable gas. Instruments are approved for different groups of gases with various ease of ignition. Where it is not possible to produce an instrument whose circuitry is intrinsically safe, it is generally enclosed in a flameproof box. This is a casing which is put together in such a way that where joints occur the flanges have a minimum gap which will quench any explosion travelling through it.

Where an instrument uses hot filaments as sensors, then under normal conditions these should not ignite the gas which is being measured. Moreover it is a general requirement that such filaments shall be protected by a suitable flametrap. Most instruments measuring by diffusion will have a sintered metal disc between the cavity containing the elements and the outside atmosphere. This sintered disc is generally an effective flametrap against any gas. Instruments which sample by aspiration will again have a suitable flametrap downstream of the cavity containing the sensors. These flametraps are tested by filling the instruments with flammable

Table 2.7 Range of environmental variables

Variable	Range
Temperature	−5 to 40 °C (−40 to 75 °C for heads)
Pressure	920 mbar to 1150 mbar
Humidity	5 to 90% RH at 40 °C
Air speed	0 to 6 ms^{-1}
Orientation	360° about any axis (portable instruments)
Vibration	10 to 500 Hz

gas and igniting the gas inside the element cavity. The flametraps should be such that no explosion is transmitted to the outside atmosphere.

Many standards will require that the case of the instrument shall be constructed of a material which will not become charged by static electricity, so that discharges to earth cannot produce an igniting spark. Materials with a surface resistance of less than 10^8 ohms per centimetre are generally suitable. In addition, it is often a requirement that the case of the instrument shall not be made of light metal alloys. Cases of such metals if dropped on to rusty steel can produce an incendive spark by reaction of the alloy with the iron oxide. Hence it is required that the total concentration of aluminium, magnesium and titanium in an alloy shall not exceed 15% by weight and, where aluminium is not present, the total concentration of magnesium and titanium shall not exceed 10% by weight.

Most standards also have a minimum requirement for robustness. This will involve dropping spherical weights on to all vulnerable parts, such as indicator lights, from specified heights, and dropping the whole instrument, if it is portable, or the sensing heads, if it is not, on to concrete from a height of one metre. The instrument will be expected to function normally after such tests.

Many continuous-monitoring instruments are designed so that the central control point is in an area where flammable gas will never be present and only the sensing heads are in classified areas. In these circumstances obviously electrical requirements for flammability will not need to be met by the control point. However, faults in the control points should not interact in the heads in such a way that the heads can become a hazard.

4 Applications of instruments

Present-day flammable gas detectors measure the concentration at the location where the instrument or sensing head is sited. In effect they are point-reading devices. However, when a leakage of flammable gas occurs, the gas at the point of emission is generally well above the LEL and in many cases has a concentration of about 100% v/v. This must automatically mean that when a leakage occurs, a cloud of gas exists which is capable of supporting a flame. The size of this cloud is, to some extent, a measure of the hazard and will depend on factors such as the rate of emission of the gas and the atmospheric conditions around the leak, especially airflow and turbulence in the airflow. Mixing by turbulence is very much faster than mixing by diffusion and the physics which controls the size of a flammable gas cloud is complex. However, since the object of monitoring these concentrations is to limit the size of the hazard and give an alarm so that rectifying measures can be taken, it is obvious that the position of the measuring point or points relative to the point at which leakage occurs, is very important. Unless the flammable gas detector is right on top of the source of the leak, then an explosive hazard must exist when any flammable gas is measured by it.

What then does the concentration measured by the detector mean in terms of the hazard? Some simple cases are illustrated in Fig. 2.24. If action is taken when the concentration level at a measuring point is a certain fraction of the LEL, then a

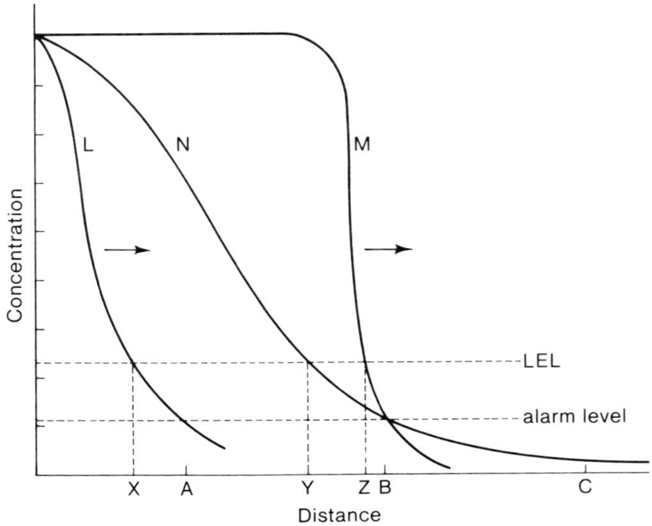

Fig. 2.24 Gas concentrations downwind of leak.

number of conditions can occur, depending on where the measuring point is relative to the source of the leak and on the rate of mixing of the gas downwind to the leak. If atmospheric conditions are stagnant and little mixing occurs, then a gas cloud will develop with a fringe in which high concentration gradients exist (curve M). This concentration front will be moving away from the source. Thus, when the concentration at point B reaches the alarm level, the hazard which exists can be represented by the area under curve M to the right of point Z on the abscissa. At measuring point C, the flammable gas concentration will be virtually zero, even though a sizeable hazard exists.

If more rapid mixing is occurring due to turbulence etc., then eventually a steady state concentration gradient will be set up downwind of the leak. This could be described by curve N. Under these conditions, an alarm of a concentration at point B will be an indication of a hazard represented by the area under curve N to the left of point Y on the abscissa. This is a smaller hazard than that represented by the same concentration level at point B where little mixing is occurring. Where such mixing is occurring, measurable concentrations of flammable gas will be detected at point C, much further downwind of the leak than where little mixing is occurring. Shortly after a leak has started under good mixing conditions, concentration gradients downwind could be represented by curve L, which in time will approach curve N. A detector at point A will alarm in conditions where the hazard is represented by the area under curve L to the left of point X on the abscissa. This hazard is obviously much smaller than that represented by the same concentration measured at point B downwind.

The conditions described above are obviously very much simplified. In practice considerable fluctuations of concentrations will occur with time and the size of these will depend on the position of the measuring point with respect to the source and the

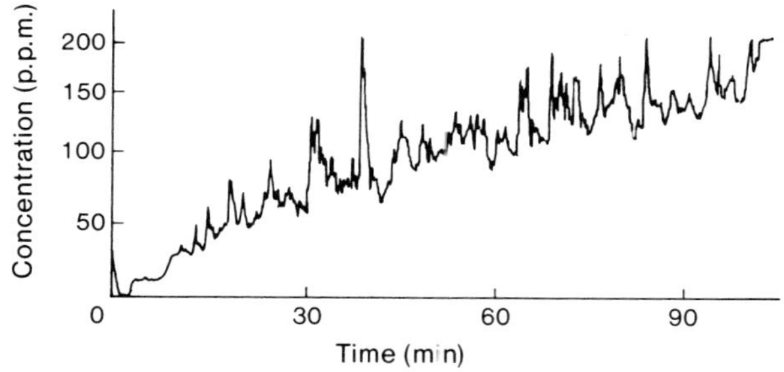

Fig. 2.25 Gas concentrations at a point downwind of leak.

axis of the resulting plume of gas.[35,36] There is little information available on the time variation of vapour concentrations approaching the LEL which result from a continuous leak of vapour. Typical variations with time at concentrations well below the lower explosive limit are shown in Fig. 2.25, measured inside a building downstream of a steady emission of styrene vapour. One might well expect similar behaviour at higher concentrations from more substantial emissions.

Thus it can be said in general terms that a concentration of flammable gas measured by a detector is, by itself, not an adequate measure of the hazard which exists. In fact, the siting of the detector relative to the leak and the mixing conditions are more important factors governing the size of the hazard than the actual concentration measured. Because flammable gas sensors are point-measuring devices, the siting of them is an important factor governing the effectiveness of their use as hazard indicators.

Since, for an explosion to occur, a flammable gas must be present with an igniting source, instead of siting gas detectors downstream of possible leakage sources they can be placed upwind of potential igniting sources. Such use does mean that, when an alarm is given, a substantial cloud of flammable gas may already exist and if sensors are used in this way, rapid or automatic removal of the igniting source is required, e.g. switch off electrical power. Where possible sources of leakages are known and relatively few igniting sources exist, this may be a more effective way of using detectors. Where few possible leakage sources exist, however, it is more desirable to measure concentrations as close to them as possible. In any case, measurements should be made at points between sources of leakage and igniting sources.

Two general types of conditions occur where flammable gases may be encountered. These are: (i) in normal operations flammable gas is not present and virtually fresh air conditions exist; (ii) in normal operations flammable gas is present and may be produced as part of the normal working environment.

In the first condition, the alarm level can be set as low as is practical taking into account the variations in zero signal normally present in the detector. Thus, if the zero signal can be expected to drift over a range of 0.1% LEL, an alarm level set at 1% LEL could be used.

In the second type of situation, the layout of the plant will have to be carefully designed so that possible leakage sources and igniting sources are well separated.

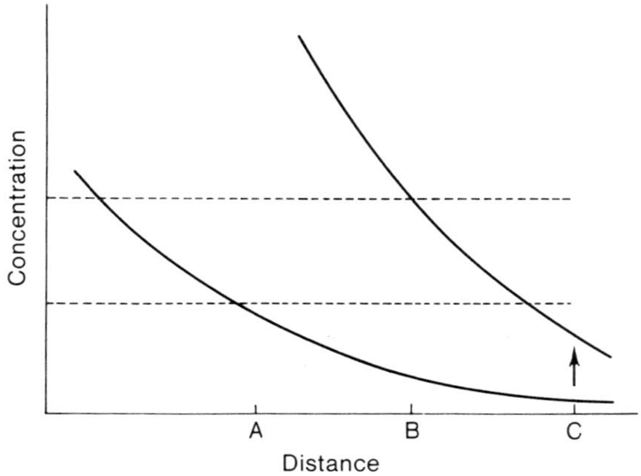

Fig. 2.26 Gas concentrations downwind of leak.

Under these conditions any leakage of gas is diluted to concentrations well below the LEL before it comes into contact with any igniting source and good ventilation conditions are essential. In these situations, changes in gas concentrations away from that normally encountered in safe operating conditions are a better measurement of hazard than an absolute measure of the concentration level itself. This is illustrated in Fig. 2.26. If a measurement of concentration is made at point A, which is between the igniting source point B and the leakage source, then the concentration at point B is always below the LEL, and a hazard will not exist. However, if the measuring point is at some position C, which is not between the leakage and the igniting source, then a change in concentration at point C even to some relatively low value may well indicate serious hazard conditions existing around the igniting source. Indeed, a rapidly increasing concentration, even at relatively low concentration levels, may well be an indication of a serious leakage of gas occurring in another part of the plant.

It is not possible to cover every factor relating to the use of flammable gas instruments in industry. However, some important considerations relating to major applications of instruments are given in Table 2.8.

5 Future developments

In many respects, high accuracy is not of major importance in flammable gas detectors but high reliability is. The limitations in the use of various systems are related to their reliability. Most of the effort being expended in developing flammable gas detector methods is aimed at improving reliability.

The main reliability problem of catalytic sensors is the possible effect of poisons on the catalyst. Ways in which the resistance to poisoning of these elements is being investigated have been described. There is little doubt that if a completely poison-resistant catalytic detector could be developed, it would meet the vast majority of flammable gas detector applications. However, the very nature of the catalytic

Table 2.8

Industrial application	*Factors relating to gas emission*	*Factors relating to gas detectors*
(*a*) Coal mining	(1) Gas is primarily one component (methane)	(1) Charcoal filters will give adequate protection to catalytic systems
	(2) It is lighter than air and collects at roof	(2) Sensors for continuous monitors should be mounted near roof
	(3) It can accumulate in pockets, in crevices, etc	(3) Portable instruments are required for inspection, preferably non-ambiguous
	(4) Worker, electrical equipment and machine are well defined ignition sources	(4) Portable continuous monitors can be used. Sensors can be mounted near igniting sources
	(5) Ventilation is good and gas is mixed by some point of coal district	(5) Concentrations at this point are a measure of general hazard on face
	(6) Gas is always present	(6) Changes in concentration at 5 are important
(*b*) Civil engineering tunnelling	(1) Factors 1, 2, 3, 4 and 5 of (*a*) apply	(1) Factors 1, 2, 3, 4 and 5 of (*a*) apply
(*c*) Sewage works	(1) Factors 1, 2 and 3 of (*a*) apply	(1) Factors 1 and 2 of (*a*) apply
	(2) Gas is not normally present in buildings	(2) Low concentration level alarms could be used
(*d*) Gas supply industry	(1) Gas is primarily one component (methane or hydrogen)	(1) Charcoal filters will give adequate protection to catalytic systems
	(2) Gas is lighter than air and can collect at roof	(2) Continuous monitor sensors should be mounted at roof
	(3) Gas is not normally present in general atmosphere of buildings	(3) Low concentration level continuous alarms could be used
	(4) Gas can accumulate in pockets as well as get into general atmosphere of building and is measured in 'bar holes'	(4) Portable instruments with high ($\leq$100% v/v), medium ($\leq$100% LEL) and low ($\leq$10% LEL) ranges are needed
(*e*) Liquefied petroleum gas (LPG)	(1) Gas is heavier than air	(1) Continuous monitors should be mounted near floor and, if possible, in a bund under storage tanks
	(2) Gas has high molecular weight	(2) Charcoal filters cannot be used to protect catalytic elements from poisons
	(3) Catalyst poisons are rarely present	(3) Unprotected catalytic elements can be used

Industrial application	*Factors relating to gas emission*	*Factors relating to gas detectors*
(*f*) Petroleum supplies	(1) Factors 1 and 2 of (*e*) apply	(1) Factors 1 and 2 of (*e*) apply
	(2) Tetraethyl lead is present as a poison	(2) Continuous catalytic monitors should be poison resistant or not used. Portable instruments require frequent recalibration
(*g*) Chemical industry	(1) Wide variety of gases may be present	(1) Detectors should be calibrated with gas to be detected. If this is not known, then errors of up to 50% can occur with catalytic sensors
	(2) Poisons may be present	(2) If poisons are present, infra-red sensors or catalysts proved to be resistant to the identified poisons should be used
	(3) Gases may be heavier or lighter than air	(3) Siting of sensing heads relative to leak is important
(*h*) Oil industry	(1) Factors 1, 2 and 3 of (*g*) apply	(1) Factors 1, 2 and 3 of (*g*) apply
	(2) Sulphur containing gases can be present	(2) Long term poisoning effects of sulphur gases on catalytic elements have not yet been studied
	(3) Extremes of environment are present on oil rigs	(3) Effects of temperature on calibration must be taken into account
	(4) In tankers low oxygen concentration will be present	(4) Gas to catalytic sensors must have air added. Infra-red sensors may be more suitable
	(5) Oil rig or tanker will be at sea	(5) Salt spray corrosion protection is necessary
(*i*) Paint drying ovens	(1) Factor 1 of (*g*) applies	(1) Factor 1 of (*g*) applies and charcoal filters cannot be used on catalytic systems
	(2) High temperatures are used	(2) Gas can condense if extracted out of plant to sensor
	(3) Silicones are present	(3) Poison-resistant catalytic element required or infra-red sensors are required. Instruments using a flame may be applicable

processes makes it unlikely that a detector resistant to all possible poisons could be developed. It seems more likely that a range of detectors will result, each capable of providing protection against a given type or range of poisons.

It will be apparent from the discussion in Section 2.4, that a sensitive low concentration continuous gas monitor could be very useful. Such a device would have an alarm level at say 1% of the lower explosive limit and would be applicable in those areas where gas is not normally present. A device with this sort of alarm level would need to have a long term zero drift and sensitivity drift of less than 0.1% LEL. A simple cheap device with a high degree of reliability could have wide applications. Semiconductor sensors offer this sort of performance and may give rise to such a device.

The difficulties of giving reliable measurement of flammable gas mixtures where the relative proportions of the different components can change have been described. To some extent, this difficulty could be overcome by the development of catalytic devices in which the signal is not diffusion controlled. In such a device, the relationship between LEL and heat of oxidation would be more closely followed and the signal would be less dependent on the constitution of the gas.

An alternative approach to this problem might be to utilise the tremendous amount of information available in an infra-red spectrum of a gas sample. An automatic dispersive infra-red analyser, coupled to a microprocessor, which is capable of both processing the spectrum and relating it to stored reference spectra, would be capable of analysing the different components in the gas mixture. Such an instrument would have to operate intermittently on account of the time taken to scan the spectrum and would only be able to analyse a sample every few minutes. However, it would be an almost complete answer to the problem of estimating flammable gas concentrations in some industries.

The problems associated with present-day instruments, which in essence only measure gas concentrations at a given point, have been outlined. Much more work is necessary on the physics of gas dispersion from given leakage points at given rates of leakage so that reliable codes of practice can be drawn up for the siting of sampling points.

If reliable siting of sampling points becomes possible on a routine basis and the reliability of sensors themselves improves, particularly with respect to poison resistance of catalytic instruments, it may be possible to consider gas detectors as a primary means of protection against explosive hazards. At the present time, they are only secondary or support systems but flameproof enclosures, flametrapping systems, etc. are very expensive. A gas detection system, coupled to automatic switch-off of igniting sources or, indeed, quenching or flooding the igniting source with inert gas may well be a cheaper and hence, more widely used, form of protection.

Present-day sensing systems try to obtain coverage of an area in which flammable gas may occur by suitable siting of individual measuring points, with all the limitations which have been discussed. The development of sensing systems capable of scanning for gas concentrations over a wide area, would be a logical development of flammable gas detection as a whole. LIDAR systems[37] have been developed for remote scanning of atmospheric pollution. If this type of system can be developed for industrial plant applications, it would represent a significant advance in the improvement of safety for flammable gas detectors. Simple initial experiments along these lines have involved using laser systems scanning across an area in which leakage might occur to an infra-red sensor on the opposite side of the area. Such systems obviously only detect gas concentrations along the path of the laser and are subject to interference by variations in the refractive index of the atmosphere along the laser beam which can deflect the beam away from the infra-red detector. LIDAR systems would overcome this defect. Although these systems are much more complex than present-day devices, the increase in computing capability brought about by microprocessors and the steady increase in the capability of radiation detectors, may make them possible systems for large plant monitoring in the not too distant future.

5 References

1. H. F. Coward and G. W. Jones. *US Bur. Mines Bull. 503.* US Gov. Printing Office, Washington (1952).
2. *Air Analysis in the Mining Industry.* NCB Scientific Control (1973).
3. G. Broomhead and S. R. Howarth. *SMRE Research Report 281.* HMSO (1972).
4. *Mine Gases.* National Coal Board, Industrial Training Branch (1960).
5. F. W. Pritchard and B. A. Phelps. *Colliery Engng.* **38**, 103 (1961).
6. P. Bussiere *et al. C.r. hebd. Seanc. Acad. Sci, Paris,* **243**, 1870 (1956).
7. A. R. Baker. Brit. Pat. 892 530 (1958).
8. J. Sieger. Brit. Pat. 864 239 (1958).
9. J. G. Firth. *Int. Conf. Directors Safety in Mines Res., Tokyo* (1969).
10. J. G. Firth. *Trans. Faraday Soc.* **62**, 2566 (1966).
11. A. R. Baker and J. G. Firth. *Mining Engineer.* **128**, 237 (1969).
12. J. M. Thomas and W. J. Thomas. *Introduction to Principles of Heterogeneous Catalysis.* Academic Press, New York and London (1967).
13. F. Klauer. *Glückauf,* **103**, 379 (1967).
14. T. M. Sporton. Brit. Pat. 1 447 488 (1976).
15. J. G. Firth, A. Jones and T. A. Jones. *Comb and Flame,* **21**, 303 (1973).
16. M. J. Burgess and R. V. Wheeler. *J. Chem. Soc.* **99**, 2013 (1911).
17. J. G. Firth, A. Jones and T. A. Jones. *Inst. Chem. Eng. Symp. No. 39* (1974).
18. C. N. Satterfield and T. K. Sherwood. *The Role of Diffusion in Catalysis.* Addison-Wesley (1963).
19. S. J. Gentry and A. Jones. *J. App. Chem. Biotechnol.* **28**, 727 (1978).
20. Young Ho Kim. US Pat. 3 586 486 (1971).
21. J. G. Firth and H. B. Holland. *Nature,* **212**, 1036 (1966).
22. A. Jones *et al.* Brit. Pat. Application 8 028 675/80.
23. W. E. Garner (ed.). *Chemistry of the Solid State.* Butterworths, London (1955).
24. T. A. Goodwin and P. Mark. *Progress in Surface Science* (ed. S. G. Davison). Pergamon Press, Oxford (1972).
25. N. Taguchi. Brit. Pat. 1 280 809 (1970).
26. B. Bott *et al.* Brit. Pat. 1 374 575 (1970).
27. J. G. Firth *et al.* Brit. Pat. Application 49 879/77.
28. M. L. Veingerov. *Dokl. Akad. Nauk SSSR,* **46**, 182 (1945).
29. K. V. Luft. *Z. Tech. Phys.* **24**, 97 (1943) and *Glückauf,*, **98**, 493 (1962).
30. D. W. Hill and T. Powell. *NDIR Gas Analysis.* Adam Hilger, London (1968).
31. H. Rieck. *Semiconductor Lasers.* MacDonald (1970).
32. H. Levinstein. *Analyt. Chem.* **41**, 81A (1969).
33. *Canadian Standards Ass. C22.2,* No. 152 (1976). British Standard 6020 (1981).
34. *Dept. of Trade and Industry, Testing Memorandum No. 7.* HMSO (1971).
35. I. A. Singer. *J. Air Pollution Control Ass.* **11**, 336 (1961).
36. F. Gifford. *Int. J. Air Pollution,* **3**, 253 (1960).
37. R. L. Byer. *Quantum Electron.* **7**, 147 (1975).

3 Oxygen deficiency

L. R. Cooper

1 The hazards

The necessity of an adequate supply of oxygen to support human life is well known and has been specifically recognised since the early scientific work of Scheele, Priestley and Lavoisier in the eighteenth century. In the practical sphere, the effect of oxygen deficiency or 'foul air' has been known from early times by men working in mines and going into caves. In these places oxygen has either been removed by a slow burning process or been displaced by other gases such as methane or carbon dioxide.

In the present day there are additional situations in which oxygen deficiency can occur. In sumps, heavy gases which displace oxygen may have collected. A similar condition may occur in manholes and tanks, which have been previously filled with another gas and then not sufficiently flushed out with air. These may all present a serious hazard due to lack of oxygen.

Such oxygen-deficient conditions would be revealed by a measurement of oxygen in the total atmosphere when it would be found to be below the normal concentration of 20.8%. Another class of condition which can be harmful to man is that caused by high altitudes. At elevations above sea level greater than 3300 m, respiratory difficulties are often encountered. In these situations the pressure of the atmosphere is significantly reduced (at 3300 m the atmospheric pressure is 700 mbar when the zero level pressure is 1013 mbar), although the proportion of oxygen in the atmosphere remains the same as that found at sea level, namely 20.8%.

At first sight it appears that there are two very distinct conditions under which the human body can suffer from the effects of oxygen shortage, the one in caves, sumps, etc. and the other at high altitudes. In the first set of situations the pressure is normal and the percentage of oxygen low, in the latter case the pressure is low and the percentage is normal. Medical tests have shown that there is only one basic factor common to all conditions of oxygen deficiency, this is the partial pressure of the oxygen in the atmosphere. The meaning of this will be considered further in the next section.

The onset of a shortage of oxygen can sometimes be recognised by the human body. The heavy breathing encountered at high altitudes is an example of this. Unfortunately, the reaction of a person can in no way be relied upon to warn of the onset of a dangerous condition. Fatalities in the conditions referred to are not common, but occur frequently enough to demonstrate this. It is necessary to use an instrument which can measure the degree of oxygen deficiency in order to obtain a reliable indication of a dangerous situation.

A number of instruments are available which can be used either as measuring or alarm instruments or both. They are not in general very complex to use but if they are misused, false alarms can result with the attendant loss of confidence in the instrument. With understanding and care in use, instruments for detecting oxygen deficiency can give reliable warnings and therefore increased safety.

2 The reaction of the human body to oxygen deficiency

Medical evidence shows that the partial pressure and not the percentage by volume concentration of the oxygen in the atmosphere is a measure of the effects of oxygen deficiency on the human body. Many papers and reports still quote oxygen levels in terms of percentage concentration and sometimes add the proviso that they apply at sea level only. It is the author's opinion that it would be more correct and simpler and would avoid misunderstandings to work throughout in units of partial pressure. This would avoid having to make corrections for height and other factors which cause a change in total pressure. In the remainder of this chapter, partial pressure will be the primary consideration. In order to help the reader who is not accustomed to this method of measurement, two values of oxygen level will be quoted at places where it may help. The partial pressure of oxygen will be given first and then in brackets the equivalent volume concentration at a total pressure of 1013 mbar (1 atmosphere), e.g. 190 mbar pO_2 (18.8%). It should always be remembered that the equivalent percentage concentration does not apply at any other atmospheric pressure than 1013 mbar.

2.1 Partial pressure

The partial pressure of gas is defined by Daltons Law of Partial Pressures which states that the total pressure of gas is the sum of the partial pressures of the separate gases composing it, that is the sum of the pressures which each gas would exert if it were confined alone in the volume occupied by the mixture. In a normal atmosphere consisting of oxygen and nitrogen only, therefore, a total pressure of 1000 mbar is in simple terms made up of 210 mbar from the oxygen and 790 mbar from the nitrogen. One of the difficulties in the use of partial pressure for the measurement of oxygen content is that there are very many units of pressure that are used from time to time. In 'old' British units we have the 'atmosphere' and 'feet of water', in pre-SI CGS units the 'mm of mercury' was often used, and now in SI there is the 'Pascal' which equals one Newton per square metre and the 'bar' which was used previously to the introduction of SI units and is now permitted in the SI system. The relationships between the units are:

100 kilo Pascals = 1 bar = 1000 mbar
1 atmosphere = 760 mm mercury = 1013 mbar
1 foot water = 30 mbar

It was stated earlier that the proportion by volume of oxygen in a normal atmosphere is 20.8%. This is fixed, but the partial pressure of oxygen in the open air can be stated only if the atmospheric pressure is known. This will be seen to follow from Dalton's Law of Partial Pressures since, if air is taken as a mixture of nitrogen and oxygen only, and since the volume proportion of these two is constant, the partial pressure of each must increase as the barometric pressure increases and must decrease as the barometric pressure decreases. At a barometric pressure of 1013 mbar (1 atmosphere) in the open air, the partial pressure of oxygen (pO_2) is 211 mbar and for the present purpose this will be rounded off to 210 mbar. At the lower extreme of barometric pressure in Great Britain of 925.5 mbar (lowest since 1870) the partial pressure of oxygen is 192.5 mbar, whereas at the highest barometric pressure (since 1870) the partial pressure of oxygen is 219 mbar. The range of pO_2 values is thus between 192 and 219 mbar with a 'normal' figure of 210 mbar.

2.1.1 Effect of other gases in air

In the above estimates it has been assumed that air is a simple mixture of nitrogen and oxygen only. There are however normally present also significant amounts of water vapour, argon and carbon dioxide, the presence of which reduce the partial pressure of oxygen. Water vapour and argon act similarly to nitrogen, that is their only significance is to displace the oxygen. Carbon dioxide is, however, toxic and must be considered as such in addition to the fact that its presence often means a reduction in oxygen content. The effects of mixed constituents in the atmosphere are considered in some detail by van Liere and Stickney in their book *Hypoxia.*[1]

2.1.2 Effect of height

As a first view of the effects of oxygen deficiency on man it must be remembered that some of the population of Great Britain live at a partial pressure of 192.5 mbar purely as a consequence of low atmospheric pressure. Elsewhere in the world people live at still lower pressures; for instance in Mexico City at an altitude of 2450 m, pO_2 is typically 162 mbar and at Namche Bazaar, height 4000 m, in the Himalayan Mountains in Nepal, pO_2 is typically 134 mbar. As is well known, there is an acclimitisation process whereby the human body adjusts to low pressures over a period of days but this is no help in conditions where the oxygen level drops quickly (in seconds or minutes) which are our main concern. These examples nevertheless do serve to show that values of pO_2 down to 134 mbar do not always cause serious harm to man.

2.2 Medical data

In the following paragraphs some published medical information is given on the effects on man of oxygen deficient atmospheres. Much simplification of the evidence has been inevitable and the reader is cautioned to consult the references at the end of the chapter before making detailed use of the data.

The harmful effects of oxygen on the human body cannot be simply stated as a danger level above which conditions are safe and below which harmful effects occur.

It is necessary to take into account the oxygen level, the activity being undertaken and the person involved. The effects of oxygen deficiency are gradual, ranging from a measurable slowing down of the brain's reactions at levels around 170 mbar pO_2 (17%) to possible loss of consciousness below 100 mbar pO_2 (10%).

2.2.1 *Effect of carbon dioxide*

In addition to considering the quantity of oxygen in an atmosphere it is also necessary to take into account the constituent parts of the remainder of the air in order to assess the medical hazard completely. Our chief concern here is whether there is a significant quantity of carbon dioxide or whether the make-up gas is chiefly nitrogen. Unfortunately, many processes that remove oxygen from an atmosphere produce carbon dioxide, which has a toxic level (TLV) of 0.5%.[2] This aspect, although strictly beyond the terms of this chapter, needs to be considered together with oxygen deficiency since the two often occur together. For example breathing in a confined space produces carbon dioxide and oxygen deficiency at the same time. The presence of carbon dioxide is not recognised by the human brain and increasing amounts of this gas cause a steady increase of loss of human control of the body until the subject becomes unconscious. In contrast, in an oxygen-deficient atmosphere in which the only other constituent gas is nitrogen, the subject can often recognise the abnormal situation by experiencing difficulty in breathing before the ultimate low level of oxygen is met at which unconsciousness occurs. This condition is of course the one met with at high altitudes. It is this inability of the human body and brain to recognise the oxygen deficient, excess carbon dioxide, atmosphere which makes the use of detection instruments so important.

2.2.2 *Responses to oxygen deficiency*

The responses of a normal healthy man are tabulated below from published data (see for example Stickney[1] and Patty[3]). The onset of impairment of judgement is difficult

Table 3.1

pO_2 (mbar)	*Concentration O_2 at 1013 mbar*	*Effect*
170	17%	Impairment of judgement starts to be detected
160	16%	First signs of anoxia appear
160–120	16–12%	Breathing and pulse rate increase, muscular co-ordination is slightly impaired
140–100	14–10%	Consciousness continues; emotional upsets, abnormal fatigue upon exertion, disturbed respiration, are all observed
100–60	10–6%	Nausea and vomiting, inability to move freely and loss of consciousness may occur; subject may collapse and though aware of circumstances be unable to move or cry out
<60	<6%	Convulsive movements and gasping respiration occurs; respiration stops and a few minutes later heart action ceases

to define. The value of 170 mbar has been taken as the basis of many reported effects from a large number of experiments to measure human reactions such as short term memory, brightness threshold and visual acuity.

For the present it should be noted that, apart from the special condition where fast brain reactions are required (the effects at 170 mbar pO_2 relate largely to responses of aircraft fighter pilots), there is no significant effect on normal human beings above 160 mbar pO_2. However, it should be remembered that the results do relate to the normal subject and that any bodily malfunction that impairs the respiratory and other bodily processes will inevitably result in oxygen depletion having an effect at partial pressures higher than 160 mbar pO_2. The effects on man which have been reported by different observers show a wide scatter of the oxygen levels at which a given effect is observed. This implies that due caution must be taken in applying the data.

The further implications of these data will be considered at the end of this chapter when the application of oxygen detection instruments is discussed.

3 Methods of measurement

From the foregoing considerations, the general requirement of a sensor that is to be used to determine whether the oxygen content of an environment is adequate for the human body is that the sensor should measure the partial pressure of oxygen in the range 100–260 mbar. An instrument which measures the partial pressure of oxygen and having this scale range will include in that range the lowest pressure that is of practical interest and the highest pressure that will be met under normal atmospheric conditions. The equivalent range for an instrument which measures the concentration of oxygen is 10–22%. In this case the lower end of the range is below the limit of oxygen deficiency under the atmospheric conditions normally encountered. The upper end of the range has an adequate margin, for instrumental error, above that defined by the constitution of normal atmospheres. (It should be noted at this point that the range of concentration-measuring instruments may need modification for use at total pressures other than normal atmospheric pressure. This follows from the previous paragraph where it was shown that the partial pressure and not the concentration is the parameter which relates most closely to the medical effects of oxygen deficiency.)

Other requirements of instruments are those common to most gas detection sensors. Thus variation of the ambient temperature should have no effect and the reading should also be specific to the gas, oxygen, being measured (other interfering gases to be considered are nitrogen, carbon dioxide and any other gas which has displaced oxygen, such gases often being either flammable gases or inert gases used for purging purposes).

There are many techniques currently in use for the detection and measurement of oxygen in the ranges stated. Chemical methods, electrochemical cells, paramagnetic methods and flame lamps are the techniques of most interest in the present context where we are concerned with the hazard of oxygen deficiency rather than its measurement for process control applications.

3.1 Chemical methods

It is usually necessary to use a chemical method whenever a primary absolute standard is required. A volumetric measuring apparatus is used, such as those developed by Haldane, Orsat or Bone and Wheeler. The detailed method of use is given in detail in chemical analysis textbooks such as that by Vogel.[4] The basic principle when using this type of apparatus is to measure the volume of a sample of gas initially and after the oxygen present in the sample has been selectively absorbed. Suitable absorbents are a strongly alkaline solution of pyrogallol, a solution of sodium hyposulphite and yellow phosphorus. Accurate results can be achieved with these apparatuses only with considerable care and practice.

None of the other methods of measurement to be considered here are absolute and they will therefore require a reference for calibration. Paramagnetic instruments serve well as secondary standards but the ultimate primary reference is usually a chemical assessment.

3.2 The flame lamp

The flame lamp is not a true instrument in the normally understood sense. It has no meter or digital indication, nor does it have an alarm signal at pre-set point. As will be shown, it is not very sensitive, and the minimum oxygen deficiency that can be detected is very dependent upon the skill and training of the user. There are also errors which the user cannot control. Nevertheless, its use for oxygen deficiency detection has been proved over many years, and this fact, together with its simplicity and ease of operation and maintenance, will probably ensure its continuing use for some years to come.

Originally developed by Sir Humphry Davy and others for providing a safe source of light in a potentially explosive atmosphere, it was soon realised that the flame lamp also served the function of indicating both the presence of a flammable gas and a deficiency of oxygen. The method by which it indicates a lack of oxygen follows from its operation which is the burning of a fuel, colzalene, rubber solvent spirit, on a wick in air. An absence of oxygen in the air causes extinction of the flame.

The present form of flame lamp employs controlled air inlets (see Fig. 3.1) and wind shields. These provide reasonably controlled conditions of operation. Broomhead and Brown[5] have examined the behaviour of several types of flame lamp under oxygen-deficient conditions. It was observed that each type of lamp responded differently to reduced oxygen but that, provided they were maintained correctly, different lamps of the same type responded similarly.

The flame was found to respond to the total pressure of the atmosphere from which it can be inferred that it responds at least in part to pO_2. Broomhead's other results have been quoted in terms of oxygen concentration, and this procedure will be followed here. The reason for this is that the reaction of the flame to changes in oxygen concentration when the pO_2 remains constant was not examined, and it is not therefore possible to carry out a direct translation of the results from concentration to partial pressure.

The simplest way of using a flame lamp is to observe the point at which the flame is

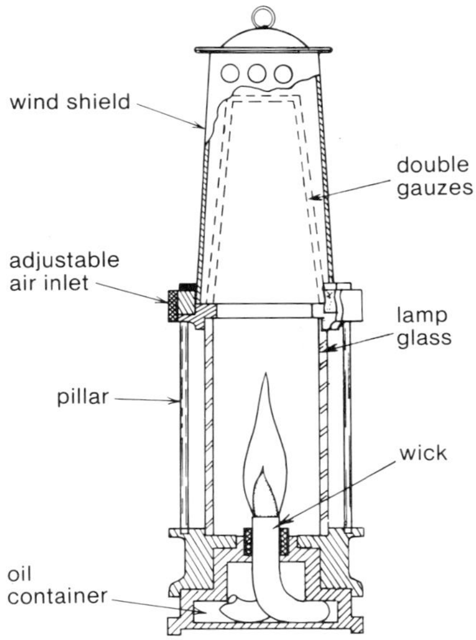

Fig. 3.1 Diagram of a flame lamp.

extinguished. For one type of lamp in common use, extinction has been observed to occur at concentrations between 15.9 and 16.6% with a mean of 16.3%. It is possible for a skilled observer to estimate higher levels of oxygen by observing the luminosity, size, shape and colour of the flame. The luminosity starts to decrease at concentrations below 19% oxygen and the height of flame starts to fall at concentrations below 18% oxygen. Colour changes are described by Broomhead[5] as 'very subtle' and not capable of being seen by all observers. Another effect which has been observed at low oxygen levels is the lifting of the flame from the wick which occurs near to the extinction point. According to Broomhead,[5] this is a reliable method of estimate by a trained observer who must know how to set the flame in fresh air (about 13 mm high) as well as to be able to recognise the lifting of the flame.

The presence of a flammable gas in an oxygen-deficient atmosphere modifies the effect described above with the exception of the level at which the flame lifts from the wick. The oxygen concentration at which extinction occurs is reduced with increasing levels of methane (and presumably of other flammable gases).

3.3 Paramagnetic methods

Paramagnetic methods of measuring oxygen depend upon the unique property of oxygen that its magnetic susceptibility is positive. In other words it is paramagnetic whereas all other common gases except nitric oxide have negative magnetic susceptibility or are diamagnetic. Thus the magnetic susceptibility of oxygen is 106×10^{-6} emu gm^{-1} whereas that of nitrogen is -0.8×10^{-6} emu gm^{-1} (Bates[13]). In a paramagnetic substance the individual atoms, ions or molecules (the latter with

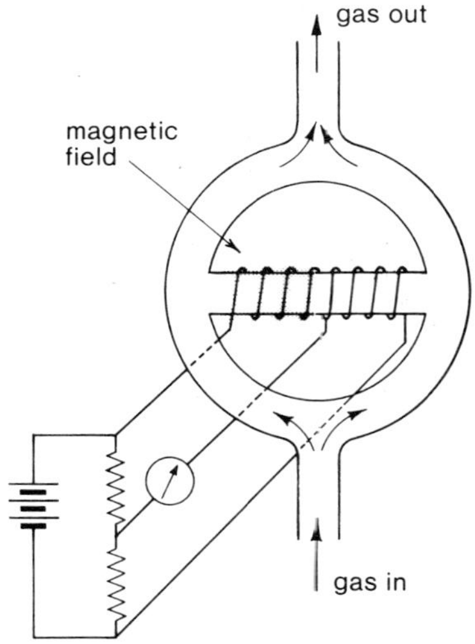

Fig. 3.2 A magnetic wind instrument (courtesy of Taylor Analytics Ltd.).

gaseous oxygen) possess a permanent magnetic dipole moment. Such dipoles line up parallel to an external magnetic field giving a net contribution to the magnetic susceptibility.

The laboratory methods of measuring the magnetic susceptibility of oxygen involve the measurement of the force on a capsule of the gas when placed in a magnetic field that has a sharp gradient. The Gouy method (see Bates[13]) is suitable for measurements on oxygen but for gases which are only weakly susceptible, more delicate methods are required. The first successful instruments employing paramagnetic methods employed a different approach; they use the 'magnetic wind' or 'thermomagnetic principle'.

3.3.1 Magnetic wind instruments

The outline of a magnetic wind instrument is shown in Fig. 3.2. A gas flow is formed by a ring having diametrically opposed inlet and outlet tubes. A connecting cross tube is sited at right-angles to the inlet and outlet tubes in one half of which there is a strong magnetic field from a permanent magnet. Two coils of a Wheatstone bridge are wound on the cross arm and the bridge is completed by two external resistances. Oxygen present in the sample gas is attracted by the magnetic field into the cross tube from the left-hand tube. In that tube it is heated by the coils of the bridge which are run at about 200 °C, and this causes a reduction in the magnetic susceptibility of the gas. The net result is that, with cool gas to the left of the field and warm gas to the right, there is a net force on the gas from left to right and a resulting net flow in that direction. This flow in turn cools the filament, the left coil more than the right coil, since the cooler gas impinges on the left-hand coil first. The bridge is therefore unbalanced and this is indicated on the meter which can, therefore, be calibrated to read oxygen content.

Portable instruments using this principle have been produced, see for example

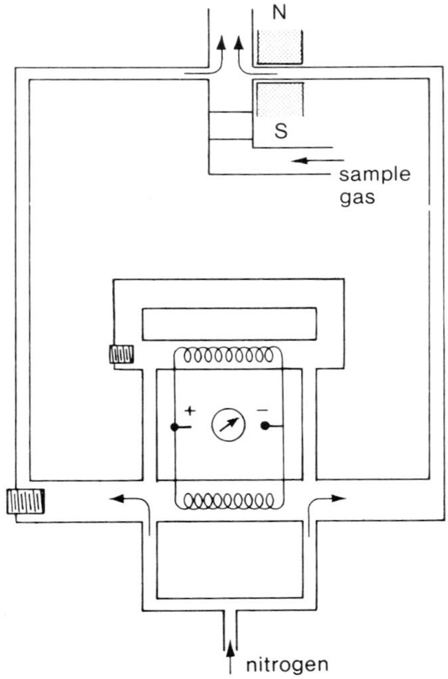

Fig. 3.3 The Quinke analyser (courtesy of Taylor Analytics Ltd.).

Luft,[6] but since they have to accommodate a powerful magnet they are rather heavy. The method has gained greater use as a fixed instrument. It is however subject to a number of errors. One arises from the fact that it is basically a thermal conductance 'katharometer' type of detector. Any changes in the gas mixed with oxygen changes the thermal balance and so gives an error signal. Furthermore some convective flow occurs which is dependent both on position and on gas composition. A third error results if a flammable gas is present which can burn at the temperature in the cross tube. The heat emitted by the combustion of the gas warms up and unbalances the coils.

3.3.2 Quinke analysers

The Quinke type of analyser shown in Fig. 3.3 overcomes some of the errors of the magnetic wind instruments. The Quinke cell has two gas flows: one of nitrogen only, the other of the sample gas to be measured. In the absence of oxygen in the sample gas, the flow paths are balanced so that there is no resulting flow of nitrogen over the coils of the Wheatstone bridge. The presence of oxygen together with the magnetic field in one nitrogen path causes a small excess of pressure to be developed in the right-hand channel. This results in a gas flow over the coil from right to left, largely in the lower bridge elements. As with the previous instrument, one element is cooled more than the other and the consequent bridge imbalance indicates the oxygen concentration in the sample gas.

Although the Quinke cell overcomes the errors caused by changes in the

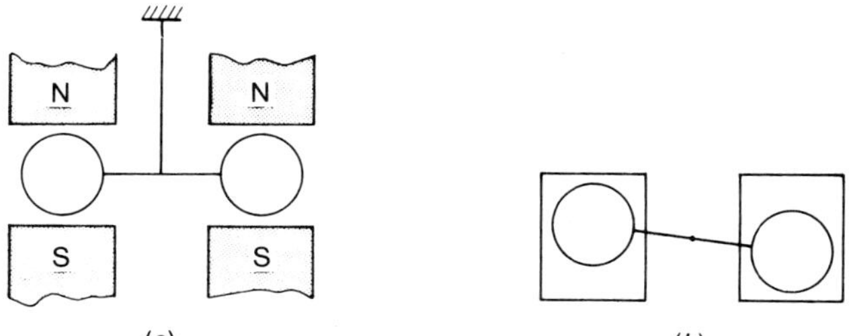

Fig. 3.4 Pauling's dumbbell system. (*a*) Side view; (*b*) plan view.

composition of the gas in the detecting coils, it has the additional complication of requiring a steady flow of nitrogen. Moreover, it is still subject to errors caused by a convective flow of gas unless close control of its attitude is maintained between calibration and use. These problems are avoided in the dumbbell system developed by Pauling[7] which is similar in principle to the basic laboratory method.

3.3.3 Pauling's dumbbell cells

Pauling's dumbbell system is shown in Fig. 3.4. The dumbell consists of two hollow spheres of diamagnetic glass filled with a gas such as nitrogen which is also diamagnetic. It is suspended by a torsion fibre in the non-uniform field of a permanent magnet. The gas to be measured surrounds the dumbbell. If oxygen is present it is drawn into the magnetic field and so displaces the spheres of the dumbbell which rotate away from the field. A steady state occurs when the force of the torsion suspension equals the displacement force. The angle of rotation is then a direct measurement of the oxygen in the sample. (It is interesting to note that in this system we have the inverse of the laboratory method from the sample point of view. Where in the laboratory method the sample is placed, in the Pauling method there is a diamagnetic gas. Where in the laboratory method the surround is diamagnetic or a vacuum, in the Pauling method there is, if oxygen is present, a paramagnetic gas. From this we can see that, whereas in the laboratory method a paramagnetic sample experiences a force into the magnetic field, the diamagnetic spheres in the Pauling system move away from the magnetic field if oxygen is present to displace it.)

In the simplest form the rotation of the dumbbell is measured directly by means of a mirror fixed on the suspension and a light and scale as is typically employed with moving-coil galvanometers. An improvement on this which avoids errors caused by a change in properties of the torsion wire involves the use of a force balance system. In this system (Fig. 3.5), coils carrying a small current are wound around the dumbbell. The current is adjusted so that the force generated on the coil by the magnetic field equals the force on the dumbbell caused by the presence of oxygen and thus no rotation of the dumbbell occurs. The current is then directly related to the amount of oxygen in the sample. Adjustment of the current may be effected either manually, observing the null position by a light spot and scale, or automatically by a servo amplifier with a light spot and photocell as the null detector.

Instruments using the Pauling dumbbell system are capable of very high accuracy

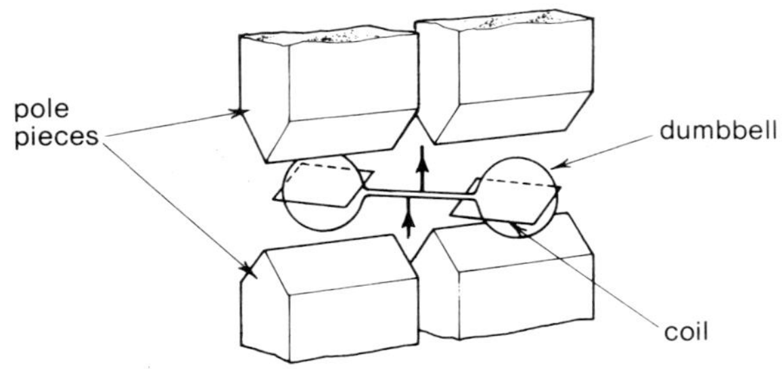

Fig. 3.5 Force balance dumbbell system (courtesy of Taylor Analytics Ltd.).

of the order of 1% of full scale reading which may be as low as 2.5% oxygen and as high as 100% oxygen. The response is dependent upon the number of molecules of oxygen present in the gas surrounding the dumbbells and therefore depends upon the partial pressure and not the concentration of oxygen.

3.4 Electrochemical cells

Electrochemical cells for the measurement of oxygen are based on the laboratory technique of polarography[9] which has been used for many years. It is only in the past ten years however that the method has been applied to portable instruments for general use. The introduction of small self-contained cells which are virtually, if not totally, leak-proof has enabled small, portable oxygen-deficiency instruments to be manufactured.

The first practical cell for the measurement of gaseous oxygen was that of Clark[10] and is illustrated in Fig. 3.6. It consists of two concentric electrodes in an insulated container covered by a membrane which is permeable to oxygen but not to a liquid electrolyte; the membrane is drawn over the end of the container flush with the electrodes, and a suitable electrolyte fills the remaining space in the container. The oxygen reaction takes place at the cathode. The cell reaction at the cathode is: $O_2+2H_2O+4e^- = 4OH^-$ and that at the anode is: $2OH^-+M = MO+H_2O+2e^-$ where M is a metal. One condition for the first reaction is that the potential of the

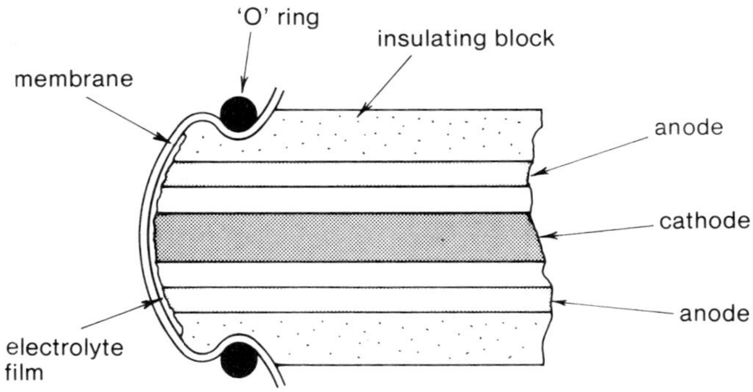

Fig. 3.6 Clark-type cell.

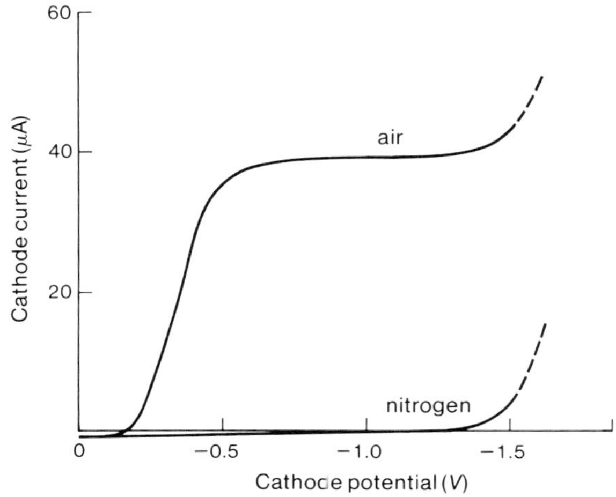

Fig. 3.7 Polarograms of oxygen cell.

cathode with respect to the electrolyte must be correct. Otherwise either at too low a potential no reaction occurs or, at too high a potential, dissociation of the OH^- ion takes place and hydrogen is emitted. This is illustrated in Fig. 3.7. In this example, if nitrogen only is present, there is negligible cell current until the cathode potential is about −1.3 V when a sharp rise of current occurs. With oxygen present the current increases steadily as the voltage changes from −0.3 V to −0.6 V and it then stays steady up to −1.3 V when there is a sharp rise. A cathode potential of 1.0 V will give optimum working for the detection of oxygen. The sharp increase in current when the voltage reaches −1.3 V is the point at which the OH^- ions start to react. (Further details relating to this reaction can be found in textbooks on polarography such as that by Crow and Westwood[9] and in the papers by Bergman and Windle.[11]) Other conditions for the cathode reaction are that the cathode material must be suitable, e.g. silver, and the rate of reaction must be low; the latter condition is normally met by limiting the rate of diffusion of oxygen through the membrane by using a suitable membrane material.

The net reaction in the cell as a whole can be seen to be a transport of oxygen from the gaseous diffusion surface through the electrolyte to form a metallic oxide at the anode, together with a current flow proportional to the amount of oxygen reacting. Since the amount of reacting oxygen is proportional to the rate of flow through membrane and electrolyte to the cathode, which in turn is proportional to the partial pressure of oxygen in the atmosphere, the current in the cell is directly related to the partial pressure of oxygen in the atmosphere adjacent to the membrane.

There are two ways of operating cells of this type, the metal used in the anode being chosen according to the mode of operation. Cells commonly called polarographic cells have an anode, sometimes called the counter electrode, which is so large that the current density is low and is made of a metal, such as silver, which makes it neutral with respect to the cathode. A potential, applied externally across the cathode and anode, is of such a value that the oxygen reaction occurs only at the cathode as shown in Fig. 3.7. As with the laboratory chemical polarographic technique, there is a small range of applied voltages over which the desired oxygen reaction occurs.

In the second mode of operation the anode is an active metal such as lead and the potential of the cathode is determined by the cathode–electrolyte–anode combination. No external voltage is required for the cell to act as a power source when oxygen is present at the cathode. This type of oxygen cell is usually called a fuel cell.

There is no fundamental difference in principle between the two modes of operation. The oxygen reaction at the cathode is the essential consideration and is the same in both cases. It may be argued for the fuel cell that no external power source is necessary and for the polarographic cell that the optimum voltage to work the cathode is easily obtained by the turn of a potentiometer. The advantages of both are finely balanced as evidenced by the fact that both types of cell are available and are being successfully used.

The types of electrochemical cell in use in most current instruments follow the basic principles of the Clark cell but with modifications made mainly to the cathode and designed to overcome a number of its limitations. One of these is the mechanical weakness of the membrane structure.

A second important limitation of the Clark cell is that gaseous oxygen must pass through not only the membrane but also the thin film of electrolyte between the membrane and the cathode. Instability may be caused by the electrolyte thickness not being constant (the diffusion rate of oxygen is controlled by both the membrane and that part of the electrolyte between the membrane and the cathode). There is another consequence of the oxygen having to diffuse through some of the electrolyte. If a step change in oxygen concentration is applied to the outer membrane surface, the time taken for the step to reach the cathode and give a signal is the same as the time taken to pass through the membrane and through the electrolyte. The longest and therefore most significant time is that in the electrolyte. For this reason Clark cells are rather slow in responding to step changes in gas concentration.

These limitations of the Clark cell have been very much reduced by the construction used in most cells now made which is shown (in outline only) in Fig. 3.8. It is necessary that the cathode in this arrangement is porous to the electrolyte and is often in the form of a paste supported on a mesh. A greater area is available for the oxygen reaction to take place and the diffusion resistance between the membrane

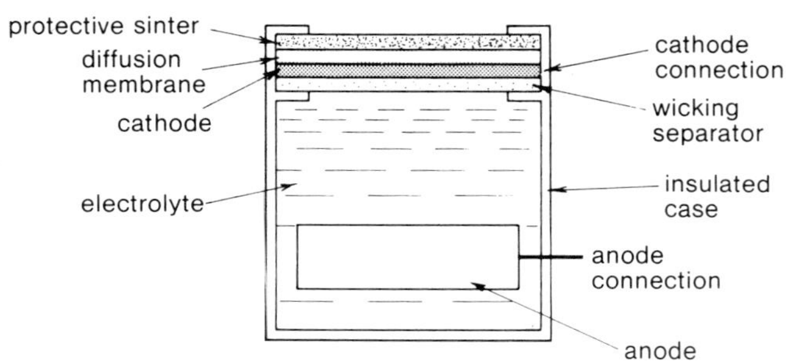

Fig. 3.8 Outline diagram of electrochemical cells.

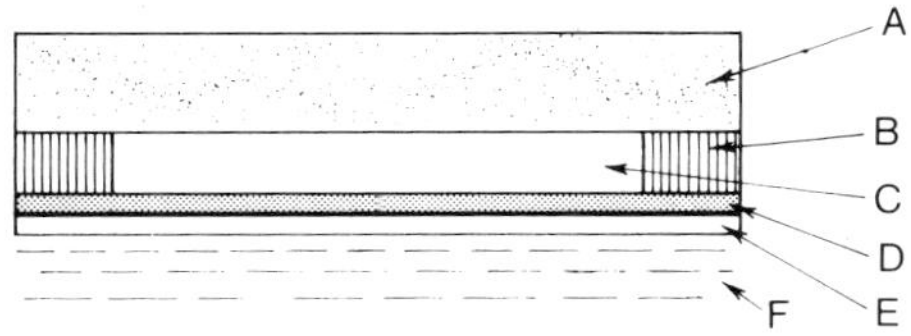

Fig. 3.9 Simplified diagram of cathode construction of Bergman cell. A, sinter disc; B, spacer; C, diffusion space; D, membrane metalised on lower surface; E, support membrane permeable to electrolyte; F, electrolyte.

and cathode is much smaller than that in the Clark cell. Response times for this type of cell are between about 10 and 30 seconds to indicate 90% of the final reading following a step change of oxygen in the ambient atmosphere.

The time response has been further reduced by a form of cathode described by Bergman and Windle.[11] The cell is shown in Fig. 3.9. In this the cathode is integral with the membrane, being formed by two thin metal films, first gold and then silver, which are sputtered on to the membrane. By this means gas diffusion through the electrolyte is almost totally eliminated and a very fast response time of 1.2 s for 90% of the final reading is quoted.

One problem encountered with all electrochemical cells which use a membrane to control the reaction rate is that the rate of diffusion of oxygen through the membrane is temperature-dependent. In the absence of any correction the signal would vary with temperature, other conditions being constant, such that the signal decreases with temperature. This is usually corrected in the electrical circuit of the cell by using a thermistor. This is a metallic oxide resistor and its resistance decreases with temperature. By choice of a suitable type of thermistor with possibly some adjustment with linear resistors, the output of the cell–thermistor combination can be made to have a negligible temperature coefficient over the range -5 to $+40\,^{\circ}$C. Sometimes, when there is good matching in the steady temperature state, errors appear under transient conditions when the temperature is changing. This is caused by the thermistor and membrane not being in good thermal contact with each other so that while in the transient state there is a temperature difference between them.

An alternative design of cathode assembly which leads to a fundamental change in operation of the cell has been proposed by Tantram.[12] Instead of a membrane being used as a diffusion barrier and limiter, a small capillary is fitted above the membrane. The size of the orifice is such that the rate at which oxygen reacts on the cathode is governed by the rate of diffusion of oxygen down the capillary, and not by its rate of diffusion through the membrane which is relatively open and serves only to contain the electrolyte. This results in two important effects. Firstly the diffusion rate through the capillary is much less temperature-dependent than that through a membrane, so that cells can be made which need no temperature compensation. The second effect is more fundamental and arises from the fact that the rate of diffusion through the capillary is dependent upon the amount of other gases present. If we are considering air, the two gases most concerned are oxygen and nitrogen. The diffusion rate of oxygen down the capillary to the cathode depends upon the amount of nitrogen present, the less nitrogen the faster the diffusion rate and conversely. The consequence of this is that the reaction rate over the range of oxygen deficiencies with which we are concerned is proportional to the oxygen concentration by volume.

Whereas both paramagnetic detectors and other electrochemical detectors considered here respond to the partial pressure of the oxygen in the sample, the cell developed by Tantram[12] responds to the volumetric concentration of oxygen in the sample. It has already been shown that there are many applications of oxygen sensor where the two measurements are equivalent. There are also some important ones where significant differences in the two measurements can occur. Some of these are considered in the next section.

4 The use of oxygen detectors

Oxygen detectors which are portable are most often required to test an atmosphere which is potentially deficient in oxygen content. This generally restricts the choice to the flame lamp and instruments containing electrochemical cells. When fixed instruments are appropriate – and they will be required mostly for process control rather than safety protection – paramagnetic instruments will often be most suitable. Although, as Broomhead and Brown[5] have pointed out, it is possible, with a number of limitations, to use the flame lamp to give an approximation of the degree of oxygen deficiency, the use of this instrument is neither accurate nor easy and this means of detection is more commonly used as a go-no-go device giving a warning when the pO_2 has fallen below a level somewhere around 170 mbar. In the ensuing discussion it will be generally assumed that an electrochemical instrument is being used.

Many applications can impose exacting requirements on the use of oxygen detectors but one use which is quite common and requires a fairly simple procedure will be considered first. It is often necessary for suspect atmospheres to be sampled before entry by passing an instrument from an atmosphere which is known to be normal, usually open air, into the unknown atmosphere. This is done by putting the detector out at arms length, or by lowering it by a rope or similar means. It is then only necessary to know if the reading falls in the unknown atmosphere, when further action, say air purging, can be undertaken before men enter. In this case the instrument can be 'calibrated' immediately prior to use and the alarm can be set a short way below the open air calibration point. The capability of the instrument to respond to an oxygen-deficient atmosphere can usually be tested prior to use by exhaling over the sampling aperture when an alarm should be given.

There are, however, many more exacting requirements where good accuracy is required. It is then necessary to know details of the behaviour of the particular instrument being used and to take account of these either in use or in the instructions for use. The characteristics which are of most concern are accuracy, long-term stability, time response and setting of alarm levels. The latter function can normally be preset by the user, who must decide the level appropriate to his application.

4.1 Choice of alarm point

The two factors to be considered in choosing the most suitable alarm point for a given application are, at the lower extreme, the medical factors as already stated and, at the upper extreme, the desirability of avoiding false alarms. In relatively simple applications where the detector is used to decide, say, whether the purging of a tank with fresh air has been completed, then a good factor of safety can be obtained by setting the alarm as close as possible to the upper point at which false alarms are not expected. This level can be well above 170 mbar if the instrument can be checked and adjusted frequently in fresh air. A more exacting application can occur in caves and parts of mines where some oxygen deficiency is normal and expected. The detector is then required to decide between the normal condition and a lower level at which action should be taken to safeguard life. It is not possible to generalise on the lower limit at which an alarm should be set in the case of such an application since, as has been seen earlier, the effect of oxygen deficiency is gradual and does not affect all bodily functions at the same time. The decision is basically a medical one and depends upon the function that a man is expected to perform in the deficient atmosphere.

4.1.1 Partial pressure instruments

In the case of a detector that measures partial pressure, the upper level of alarm setting to avoid false alarms is at first sight a simple decision as to the range of atmospheric pressure to be expected and the range of drift of the detection between calibration checks. For instance, if the instrument can be expected to be accurate to within ± 10 mbar during its period of use between calibrations, then, if the normal lower limit of atmospheric pressure is taken as 925 mbar, equivalent to a lower pO_2 of 192 mbar, the alarm point can be set at 180 mbar. This provides a low probability of false alarms and in addition the factor of safety in medical terms is high.

Some instruments which react to the partial pressure of oxygen have their indicating meters calibrated in oxygen percentage. No doubt this is done in order to make it more readily intelligible to users who are not acquainted with the concept of partial pressure, but under some conditions serious errors can result. The instrument reading in the open air changes with atmospheric pressure and it may be erroneously assumed that the calibration is drifting and so be reset to read 21%. This will not cause any harm if the instrument is being used simply as a go-no-go device at atmospheric pressure but if it is used for exacting work, dangerous errors can occur whenever the instrument is adjusted to read 21% at a low atmospheric pressure. This could lead to a positive error of around 30 mbar with the result that, if the alarm were set to the 16% reading on the assumption that this be equivalent to 160 mbar, the pO_2 would then drop as low as 130 mbar before the alarm is activated. This may constitute an unacceptable factor of safety under working conditions. For this reason, for accurate work, it is important to adjust all partial pressure instruments, however calibrated, to take account of the barometric pressure.

4.1.2 Concentration instruments

Having stated that partial pressure is the guiding physical factor when deciding upon a safe alarm point, we must now consider how to use instruments which respond to oxygen concentration and not partial pressure. In the case of the simple applications considered earlier there is no problem, since for many other applications, as will be shown, the measurement of concentration is satisfactory, provided that pressure changes are known in advance and are not excessive. For some applications where the pressures are unknown, say where air locks with appreciable pressure differences have to be passed, the conditions to be outlined do not apply and a concentration-type instrument cannot readily be used.

Firstly let us consider the common condition where a concentration instrument is to be used at atmospheric pressure only, and let us further assume that for the given working condition the instrument should never alarm at a level below 160 mbar. At the lowest expected barometric pressure of 925 mbar, where pO_2 is 192 mbar, an oxygen-deficient atmosphere at the same total pressure but with a pO_2 of 160 mbar will have an oxygen concentration of $160/192 \times 21\%$, i.e. 17.5%. If the instrument alarm is set at this level, then it can confidently be expected that the partial pressure at the alarm point will never be below 160 mbar. (If the atmospheric pressure is elevated, the equivalent pO_2 at 17.5% will be higher and the factor of safety is increased.) This method of assessment can be used for any other desired partial pressure of alarm which must of course be below the limits set by normal atmospheric changes. In the same way the alarm point can be adjusted to take account of pressure changes caused by depth or height or behind pressure seals, provided that these changes are fully predictable.

4.2 Effect of response time

The response time of an oxygen detector should be considered in relation to the use of the detector. Instructions for use will often need to specify the time for which the instrument must be put in the unknown atmosphere. For example, if a detector is to be used for testing the atmosphere in a tank prior to men entering, the detector needs to be inserted for a time which is long enough for the instrument to give a true reading. For this purpose the time to indicate to greater than 90% of the final reading is often quoted. If the exponential time constant is quoted, it must be remembered that this is the time to indicate to 63% of the final reading and that the 90% reading will be obtained within 2.3 exponential time constants.

In the above application, a man's actions can be adjusted to accommodate the time constant of the detector but this is not always the case. Consider for instance a man walking in a potentially oxygen-deficient atmosphere such as a sewer or a mine, where the exact position of a dangerous condition is not known. The detector must therefore react within the time it takes a man to proceed from a safe area into an atmosphere so deficient in oxygen that the man collapses and is unable to retrace his steps. One can postulate that under extreme conditions of a still, layered, atmosphere this could happen in a time which is very much less than a second. In practice, such a condition is very unusual and, in most cases a 90% response time of the order of five seconds is satisfactory, although it is desirable that an instrument should respond as rapidly as possible.

Intermediate requirements for the time response of detectors should be considered on their individual needs. In general the more rapid the response of the instrument the less need there will be to warn users to allow time for the instrument to respond. Since one cannot always assume that instructions are followed, all other aspects being equal, the shorter the response time, the safer will be the application of the detector.

References

1. E. J. Van Liere and J. C. Stickney. *Hypoxia.* University of Chicago Press, Chicago and London (1963).
2. HSE Guidance Note EH 15/77. *Threshold Limit Values for 1977.* HMSO (1978).
3. F. A. Patty. *Industrial Hygiene and Toxicology,* vol. 2. Interscience Publishers (1958).
4. A. I. Vogel. *A Textbook of Quantitative Inorganic Analysis*, 2nd edn. Longmans, London (1951).
5. G. Broomhead and D. K. Brown. *The Effect of Blackdamp on the Flame of a Safety Lamp.* SMRE Technical Paper, Safety in Mines Research Establishment, Sheffield, England (1975).
6. K. F. Luft. A new device to measure and warn of a deficiency of oxygen. *Glückauf,* **22**, 1137 (1967) (in German).
7. L. Pauling. *The Nature of the Chcmical Bond.* Cornell University Press, New York (1954).
8. C. W. Munday. Brit. Pat 736 778.
9. D. R. Crow and J. V. Westwood. *Polarography.* Methuen, London (1968).
10. J. C. Clark. Monitoring and Control of Blood and Tissue Oxygen Tension. *Trans. Am. Soc. Artificial Organs,* **41**, 2 (1956).
11. I. Bergman and D. A. Windle. Instruments based on polarographic sensors for the detection, recording and warning of atmospheric oxygen deficiency and the presence of pollutants such as carbon monoxide. *Ann. Occup. Hyg.* **15**, 329–37 (1972).
12. A. D. S. Tantram. Gas Sensor US Pat. 4 132 616 (Jan. 1979).
13. L. F. Bates. *Modern Magnetism.* Cambridge University Press (1939).

4 Monitoring toxic gases in the work-place

B. Miller and P. O. Kane

Preamble

The techniques used to measure toxic vapours in the field of industrial hygiene are very numerous and in principle include all the techniques of analytical chemistry. Because of this, this chapter concentrates on the methods most widely used in industrial hygiene rather than attempting to describe all the methods which could be used. The reasons for making measurements are outlined and the approach is essentially a practical one so that the methods described, with the exception of multipoint, fixed-station monitors, are within the analytical competence of the normal works laboratory rather than specialised research laboratories possessing large capital equipment resources.

1 Aims of hygiene monitoring for toxic vapours

1.1 Primary aim

The primary aim of hygiene monitoring for toxic vapours in the work-place is to provide analytical information which the workers and management can use to ensure that, as far as is reasonably practicable, no one in that work-place suffers injury or illness or a reduced quality or length of life as a result of exposure to toxic vapours associated with the work being done.

1.2 Secondary aims

(*a*) Provision of long-term analytical data for possible future epidemiological surveys; (*b*) providing analytical data to show compliance (or non-compliance) with statutory requirements; (*c*) providing information to enable improved plant operation/practices.

2 Strategy of hygiene monitoring

It is important to stress from the outset that monitoring the atmosphere of the workplace does not, in itself, protect any one; it merely provides information. The strategy of monitoring should be such that the information is provided in the place, at the time and in a way that allows action to be taken in accordance with the aims outlined in Section 1.

The strategy must take into account the nature of the toxicity of the vapour. For example, the use of a monitor which gives a retrospective time-weighted average analysis would be useless in protecting a worker on a plant from being killed by a lethal concentration of an acutely-toxic vapour. Conversely, there is no point in installing a monitor which would give an alarm and lead to complete evacuation of a plant because the long-term exposure level of a chronically-toxic vapour had been exceeded momentarily. Such a monitor would very soon be ignored, and might even be rigged to prevent the alarm sounding.

A further point to note in the strategy of hygiene monitoring is that monitoring costs money. Such costs, which, except in meeting the last aim of Section 1, have no pay-off, must be added to the cost of the product of the work-place and, eventually, paid for by the ultimate customer. In a competitive situation, this leads to the need for the monitoring to be as cost-effective as possible.

Bearing all these factors in mind, the strategy which we believe to be correct is as follows.

(1) Operations in which people might be exposed to toxic vapour will be identified. These might be one part of a complex plant, the whole plant operation, a scheduled maintenance operation, etc.

(2) Consider the nature of the toxicity of such vapours. Are they acutely and highly toxic? Are they chronically-toxic? What are the concentration levels for which exposure is not considered to have a harmful effect? Are they or other possible vapours flammable? Can they be smelt at concentrations well below those which exhibit toxic effects?

(3) Consider the situation and nature of the operation. Is it indoors or outdoors? If indoors what is the ventilation? Is it highly-directional? If outdoors, how do the weather conditions compare with what might be expected at other times? Is the plant running normally?

(4) In the light of all these considerations, decide if it is sensible to do any monitoring. There is no point in monitoring a plant which has been shut down for a week if information is needed about concentrations of vapour during normal running. There is little point in monitoring a highly-soluble vapour on an open-structure plant during heavy rain. If it is sensible to monitor, decide if the nature of the operation is such that the people doing the monitoring need protection from exposure to the vapour themselves by taking advice from plant personnel. Decide if any instruments to be used for monitoring need to be intrinsically-safe.

(5) Having taken the appropriate precautions, a useful first step in the overall monitoring programme is a rapid survey of the appropriate area using either a portable (preferably continuously-indicating) monitor or a spot-test method giving on-the-spot results. In deciding the measuring techniques for conducting this survey,

the analyst must take into account possible interfering vapours, required speed of response of an instrumental monitor etc., concentration ranges to be measured, required precision of measurements and number of measurements. The aims of this survey should be to establish the extent of the problem of possible exposure of people to toxic vapours in this location. If there is such a problem, then to indicate actions that need to be taken to minimise the problem. For example, this preliminary survey might detect adventitious leaks, highlight a particular operation which releases vapour to atmosphere to an unnecessary degree, indicate the need for improved ventilation in various areas, etc. If the rapid survey indicates that in normal working conditions the situation is safe then an air sampling strategy can be determined using the considerations described in Chapter 6 and in part in Chapter 2. The preliminary survey may, however, indicate that dangerous amounts of toxic vapours or gases are present at some locations at various times. Steps will then have to be taken to modify the production process, or install local ventilation systems, change some chemicals in the process for less toxic or more easily managed ones, etc., so that when all reasonably practicable measures have been taken, then in normal working conditions the atmospheric vapour concentrations are well below those at which harmful effects can occur. In the unusual circumstances where this situation cannot be achieved, then, either the plant would have to shut, or the people on the relevant parts of it would have to wear breathing apparatus at all times.

Assuming that the normal working conditions can be made satisfactory, then the plant management and analysts would then agree a schedule of monitoring to meet the aims of Section 1. This schedule would take into consideration such factors as the toxicity of the vapour, the concentrations which had been found in the various surveys, any identified plant components or operations which had been highlighted as potential sources of atmospheric contamination and will generally include measurements made at particular times and or continuously at various locations, and measurements made with devices mounted continuously on individual workers. The latter types of measurements are dealt with in Chapter 5. This chapter will concentrate on the techniques used for making measurements at particular locations. They include short term measuring techniques, portable continuous monitors/leak testers and single or multipoint fixed monitors.

3 Methods of hygiene monitoring

3.1 Short term measurements

These are sometimes called spot tests which is, strictly speaking, an anomalous term, implying as it does, a measurement at a given place and time, or, in other words, an instantaneous measurement. In practice, of course, all measurements or sample collections must take a finite time and, consequently, the duration of a spot test is somewhat vague. The essential feature is that this duration is small compared with a working shift of eight hours so, in practice, a spot test might vary in duration from less than 1 s to 20 min ($<5\%$ of an 8-h shift). In addition to the spot test methods given in this section, most of the portable, continuous monitors described in Section 3.2 can, in principle, be used as spot test instruments, averaging the reading over short periods of time either mathematically, electrically or by eye. Spot tests can be sub-divided into two broad classes: those giving an on-the-spot analysis and those involving sample collection for subsequent laboratory analysis by a non-portable technique.

3.1.1 On-the-spot analysers

Indicator tubes

The most common on-the-spot analyser is the indicator tube, often called a detector tube, in which a hand-pump draws a known volume of atmosphere through a tube containing one or more reactive chemicals, usually supported on an adsorption type material such as silica gel. The product of the reaction between the toxic vapour of interest and the reagent(s) in the tube is coloured and the concentration of the toxic vapour is indicated either by the length of a coloured stain or by its depth of colour.

Detector tubes can be used to measure a wide range of compounds (about 100) both inorganic (such as NO_2, CO and CO_2) and organic (such as alcohols, esters and hydrocarbons), with a precision and readability ranging from 5 to 30% depending on the reaction principle. They can achieve the same order of accuracy provided interfering gases are absent or can be eliminated.

In 1976 British Standard BS5343 *Gas detector tubes*[1] was issued and the user of such tubes should read it. The standard sets out minimum requirements for accuracy, manufacturers' instructions, shelf life and marking, and use and calibration of aspirating pumps. These can be summarised as follows:

(1) The aspiration pump should maintain an accuracy of $\pm 10\%$ in sampled volume.

(2) The tube should be within specification in all humidity conditions. Where this cannot be achieved, the conditions in which the tube can be used should be given.

(3) The tube should be within specification over the temperature range 5–35 °C. Where this cannot be met, the conditions for use should be specified.

(4) The chemical reactions in the tube should be described by the manufacturers and possible interferences given, especially where these could give rise to low results.

(5) The tube should be within the specification after storage for at least one year at temperatures up to 30 °C unless stated otherwise by the manufacturer.

All the manufacturers calibrate their detector tubes in the factory and normally stamp the boxes with an expiry date giving a shelf life of between one and two years and the user must discard the tubes once this date is reached because ageing of the reagents in the tube due to slow reactions, etc. between them, can reduce the performance of the tube. We have over the past ten years checked the calibration of batches of the tubes we use (about seventy different types) and with a very small number of exceptions, found their performance within the limits specified by the manufacturer.

It is essential that the manufacturer's instructions are followed when using indicator tubes and the user should always test the aspirating pump for leaks before each test as a leaking pump will cause erroneously low results. To test for leaks insert an unopened tube into the pump and execute a pump stroke. In the case of bellow-pumps the pump is airtight if the bellows has not expanded again completely after 10 min and in the case of piston pumps if the piston returns to within 5% of the zero mark on release after 1 min.

Unless otherwise specified by the manufacturer, indicator tubes should only be used to measure the gas(es) named on the packet. If the tube is used to measure another substance, then care must be taken to apply the proper corrections supplied by the manufacturer to the calibration scales.

When analysing an atmosphere for a particular compound the user is likely to be sampling a mixture of gases and in some cases there will be interference in the determination which could affect the accuracy of the test. To overcome this problem the manufacturer, whenever possible, incorporates prelayers to remove interfering compounds but will generally state either on the instruction leaflet or in a handbook[16,17] the effect of any known interfering compounds on a particular test. The user should check the effect of any known or suspected interferant, either by seeking the advice of the manufacturer or checking the effect of known concentrations of the interferant. Normally interferants tend to give high results which are fail safe but there is always the danger of erroneously low results particularly if the interferant reacts differently either with the indicating layer or with pre-reaction layer. For example in the Draeger vinyl chloride 1/a tube, the vinyl chloride is first oxidised by an acid permanganate layer to chlorine and this then reacts with an *o*-tolidine layer to give a stain. Other unsaturated chlorinated compounds will interfere, reacting like vinyl chloride to give a falsely-high analysis, whereas non-chlorinated unsaturated materials can use up the permanganate partially, leading to falsely-low analyses.

The measuring range of some detector tubes can be extended to higher concentrations by decreasing the number of pump-strokes or to lower concentrations by increasing this number. To do the latter the operator would find it extremely arduous executing, say, 100 pump strokes and for this reason most manufacturers supply an automatic, powered pump. In some cases the required volume of sample can be drawn through the tube using a water pump or electrically operated pump but care must be taken to draw the sample through at the correct flow-rate or results may be inaccurate. When the number of pump strokes is reduced to measure high concentrations then in most cases further pump strokes in clean air will be needed to desorb any gas adsorbed on prelayers or on the indicating layer itself.

When sampling inaccessible points the indicator tube can be used with an extension hose (3 m long). The indicator tube is placed directly at the sampling point and the pump operated at a convenient distance away. As the indicator tube is fitted at the intake opening of the tube, the volume of the connecting tube does not need to be taken into consideration in the measurement. The disadvantage of this technique is that the indicator tube cannot be observed during the measurement.

Indicator tubes can be used to sample hot gases in vents (using a hot air probe) or motor vehicle exhausts using a suitable exhaust gas probe. Indicator tubes are usually calibrated by the manufacturer under laboratory conditions when temperature is about 20–25 °C, the humidity 40–60% RH and the pressure about 1000 mbar. However, tubes can be used to measure under extreme conditions[14] and corrections applied.

Although the basic concept of the colorimetric tube is simple, the underlying mechanism whereby the contaminant produces a stain in the tube can be extremely complex. A number of chemical reactions are used in such tubes[16] and these include (*a*) rapid reactions which are independent of the flow rate through the tube; (*b*) rapid reactions with secondary reactions which may depend on the flow rate through the tube; (*c*) consecutive reactions in which the vapour reacts with one reagent to produce products that are detected by reaction with a second reagent; (*d*) reactions where a large excess of reagent is necessary. For tubes involving reactions (*c*) and (*d*), flow rate is often critical because the reaction rate with the reagent can be slow. It has been mentioned earlier, that in addition some tubes involve the use of a pre-cleansing layer of a reagent which is intended to remove possible interfering substances.

When a particular reaction has been chosen for a tube design then the following factors have to be considered or controlled in the manufacture of indicator tubes:

(1) The uniformity of distribution of the reagent amongst inert granular carrier. The carrier is generally silica gel, but alumina, pumice, porous glass and alumino silicates have been used. Impurities in the carrier must be removed since they may interfere with the reagent, affecting shelf life, etc. or with the reaction of the reagent with the vapour. In addition, the carrier should not be one on which the vapour is strongly adsorbed, preventing its reaction with the reagent.

(2) The specificity of the reagents to the vapour to be tested.

(3) The slow chemical interaction of the reagents with the support and each other.

(4) The uniformity of the bore of the tube and reproducibility of the scale on the tube.

(5) The uniformity and reproducibility of the size and porosity of the granules used as the support.

(6) The packing density of the granules.

Considerations of the generalised form of the reactions in indicator tubes have been made by Saltzman.[23–25] Where the reaction rate between the contaminant and the reagent is very rapid, then simple relations between the concentration of the vapour and the stain length result.

$$L = KCV$$

where L = length of stain, C = concentration of contaminant vapour, V = volume sampled, and K is a constant.

However, in general, the rate of reaction between the vapour and the reagent is not very rapid and kinetic processes dominate the stain length. Hence, the stain length is determined by the rate of reaction between gas and solid or by the rate of adsorption on to the solid. In such conditions, it can be shown that

$$L/H = \log(CV) + \log(\mathrm{K/H})$$

where H = mass transfer proportionality factor and is proportional to flow rate.

Thus, a logarithmic relationship exists between stain length and concentration in most reagent tubes provided H and V are constant.

The above considerations emphasise the need when using reagent tubes to ensure that flow rate through the tube is maintained within the manufactured tolerances for the calibration to apply.

Indicator tubes thus provide a simple, rapid, inexpensive and convenient means of testing for toxic gases and vapours. When used with a hand pump the method is safe to use in explosive atmospheres but this may not be the case when automatic pumps are used for multi-pump stroke tests. It should be determined in such cases, whether or not the pump is certified for use in potentially explosive atmospheres (see Chapter 2).

As with all instruments the users of tubes should be trained in their use, preferably by using them to measure known concentrations.

Test papers

Test papers provide another means of on-the-spot analysis but do not cover as wide a range of compounds as indicator tubes. Test papers are impregnated with reagents specific in response to the toxic substance to be measured. When the paper is exposed to the toxic substance a chemical stain develops with an intensity which is related to the concentration of the toxic substance.

The atmosphere to be sampled is drawn through the paper using a hand pump or a portable automatic sampling unit, the volume sampled depending on the toxic substance being measured and the sensitivity of the reaction.

The stain produced is either read with an optical reader to give a direct concentration or compared by eye with a standard colour chart. (Accuracies are about $\pm 10\%$ and $\pm 25\%$ respectively.) Some of the methods used by the UK Health and Safety Executive are based on this system and booklets on specific applications such as hydrogen sulphide and phosgene are available from HM Stationery Office.[10]

Test papers are available for the determination of hydrogen sulphide, chlorine, phosgene, toluene di-isocyanate, vinyl chloride, nitrous fumes, ammonia, formaldehyde, hydrogen chloride, sulphur dioxide, hydrazine, toluene diamine, dinitrotoluene and organic isocyanates. Equipment such as optical readers, automatic samplers, colour comparison charts, paper holders and booklets of test papers are commercially available.

The test paper system described above gives a quantitative analysis and the test takes between 5 and 20 min. A rapid, semiquantitative test can be carried out using 'free' paper-tests, that is an impregnated paper is exposed to an atmosphere and a result obtained in seconds. The papers used are usually impregnated with reagents immediately before the test, exposed by hand, and the concentration estimated from the intensity of the stain produced. Such systems are available for compounds such as hydrogen cyanide[11] and hydrogen sulphide.

There are many colorimetric reactions applicable to test papers which have the versatility arising from the use of chemical reactions, while eliminating wet chemical analysis methods. The systems available do not cover the range of applications offered by indicator tube systems but in some cases can be more specific. If test papers are to be used with portable sampling equipment then care should be taken with regard to the need for intrinsically safe electrical equipment at some sampling locations. When 'free' paper tests are made there is no need to consider intrinsic safety.

Portable gas chromatographs

Portable gas chromatographs are extremely useful for direct on-the-spot readings and analyses giving rapid results and, if needed, rapid surveys can be made. One convenient instrument is the Century Systems Organic Vapour Analyser (OVA) which is a completely portable hand-held instrument which is basically a flame ionisation detector (see Section 3.2.7) giving continuous read-out. The OVA, which weighs 5.5 kg, has a direct meter read out which can have logarithmic (1–10 000 p.p.m. v/v) or linear scales (three scales 0–10, 0–100, 0–1000 p.p.m. v/v) to give total organics. A chromatographic column attachment is available to enable an on-the-spot analysis of a mixture of toxic gases. Standard columns are available covering a wide range of separations and the instrument can be fitted with a light weight recorder.

The instrument can also be fitted with a septum allowing direct on-line syringe injections of samples eliminating the need to return samples to the laboratory for analysis. Some versions of the instrument can be taken into flammable areas as they are intrinsically safe (BASEEFA Certificate 76002/B). Calibration in either total organic or GC mode is needed using standard atmospheres (see Chapter 7) of the gases to be checked. Zero checking in the field can be done either by fitting the inlet with a suitable adsorbent tube such as charcoal or by removing the instrument to a 'clean' area.

Alternative portable instruments are available for field use, which can be fitted with a range of detectors, including flame ionisation, thermal conductivity, flame photometric, electron capture and flame ionisation–catalytic converter. Samples are introduced either by syringe or through a gas sampling loop. Detection limit will depend upon the type of detector used, for example with a flame ionisation detector less than 0.05 p.p.m. propane can be found in a 1 cm^3 sample but with an electron capture detector (Section 3.2.8) less than 0.02 p.p.b. carbon tetrachloride can be detected in the same volume. The user must decide which suits his needs best with regard to cost, portability, total organics or separations, versatility and intrinsic safety.

It must be remembered that, since gas chromatography basically involves separation of the components of a mixture by injecting a plug of sample into a moving gas phase (helium, argon, nitrogen, etc.) passing through a column of granular solid often impregnated with a non-volatile liquid, the analysis of the atmosphere is not continuous. Separation of components[20] is due to their different partition coefficients between the gas and solid or liquid phase and the elution of a particular component from the end of the column into the detector can take place many minutes after its injection into the chromatograph. Therefore, although the detector used on the chromatograph generally has a rapid response (< 1 s) to a change in gas concentration and can therefore be used to measure continuously changes in concentration, a portable chromatograph is essentially an analytical instrument which can be taken to a location to provide an analysis of the components of an atmosphere and their concentrations at a given time. Continuous operation of a chromatograph will give intermittent measurements of the atmosphere.

3.1.2 Spot samples for subsequent analysis in the laboratory

Methods for field collection of samples are well documented.[2–4,18] The ones described here are those which are regularly used. There are certain precautions to be observed when using collection devices in which samples will be stored or transported prior to analysis. The devices are usually made of glass or plastic and sometimes metal so the surfaces need to be conditioned whenever possible or care taken to avoid surface adsorption of the gases collected. Possible reaction with atmospheric moisture should also be considered. In general, the more chemically reactive a gas is, the more likely it is to react with the surface of the container. Thus, nitrogen dioxide will rapidly react with the surface of glass and with water adsorbed on the surface of glass and be permanently lost for subsequent analysis. However, polar compounds, such as alcohols, will also adsorb strongly on to glass surfaces, although in some cases these can be recovered for analyses by washing the container with a solvent. Because of adsorption process, metal containers should only be used for sampling permanent gases which are chemically relatively non-reactive, e.g. carbon monoxide, carbon dioxide. When plastic sampling devices are used, it should be remembered that many plastics are solvents for organic vapours and part of the sample can be lost by dissolution of components into the plastic, followed by permeation through the plastic to the outside atmosphere. Generally, field-collected samples should be analysed as soon after collection as is possible and storage tests should be carried out prior to collection whenever possible to determine the maximum time which can elapse after sampling before analysis. Care should be taken to ensure that when taps are used, the samples will not be absorbed by grease or when plastic materials are present that adsorption does not take place. With some gases it is possible to shield taps with a small quantity of mercury.

Evacuated flask and vacuum sampler

These two devices are variants on a common theme.

The evacuated flask (Fig. 4.1) is a well established method for the collection of gas samples. The volume of sample collected will depend on the sensitivity of the analytical method to be used and the concentration of the gas(es) being sampled. Flasks may be used which are up to 5 dm^3 capacity with glass taps fitted with PTFE barrels and neck sleeves. Samples can be collected over periods up to 20 min by using a fixed length of capillary tubing (1 mm bore) to control the flow, which will, of

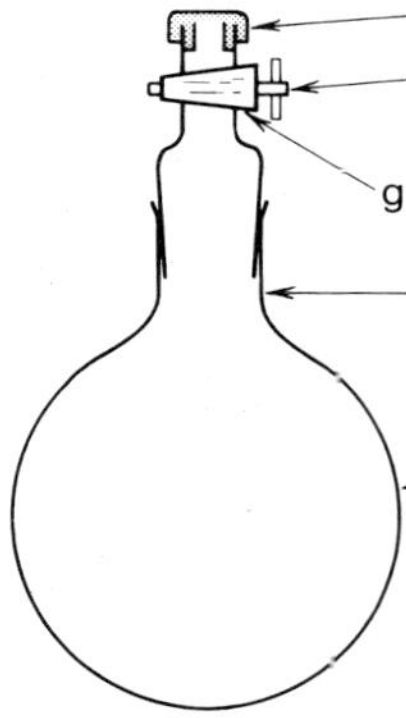

Fig. 4.1 Typical evacuated flask.

course, vary during the sampling period. The flask tap (if fitted) can be used to control the flow but we find control better with the capillary tubing. The flask can be evacuated in the laboratory or field, and before use should be checked with a vacuum gauge to ensure the residual pressure is less than 5% of atmospheric pressure. All glass flasks should be encased in plastic mesh to protect the user should the flask implode during evacuation or use. The vacuum gauge should also be used at the end of the sampling period to check if atmospheric pressure has been reached. If the flask is not filled to atmospheric pressure then either clean air must be introduced and the dilution factor calculated or the pressure in the flask noted and a correction factor applied to the concentration of toxic gas determined. If the sample collected is to be analysed chemically then reagents should normally be added to the flask either before evacuation or immediately after collection to avoid adsorption problems. For subsequent gas chromatographic analysis the flask can be fitted with a septum usually isolable from the flask by a tap, and samples removed by a syringe fitted with a long needle to pass through the bore of the tap.

Vacuum samplers are a recent innovation and have been developed for spot-sampling, breath sampling and personal monitoring.[5–7] The samplers are, as the name implies, first evacuated and the sample collected by either pressing a button to get a grab sample or by using a control valve or limiting orifice to sample over any suitable period. The advantage of a critical orifice as a control of sampling rate, is that the velocity of gas through the orifice is sonic so that the sampling rate is independent of the pressure inside the can over a large range of pressure difference up to about 500 mbar. The samplers are normally made of aluminium or an aluminium alloy and collect about 100 cm^3 which is adequate for a subsequent GC analysis. The sampler can be used as a 'pump' to draw a sample through an adsorbent trap or even through reagent in a bubbler. When used for direct sampling the problem of surface adsorption again arises. If the user suspects that adsorption can take place then experiments should be carried out to determine the rate of adsorption and/or the sample lost at different concentrations. Since the surface of the vessel will have a finite adsorption capacity, the concentration reduction by adsorption diminishes with increasing concentration of the gas. Unfortunately, it is generally true that the more chemically reactive gases are more physiologically active and have to be measured at low concentrations.

Syringe samples

Samples collected by gas-tight syringe are normally analysed using a sample loop by gas chromatography. The method is described in the Chemical Industries Association Manual[8] for the determination of vinyl chloride and can be applied to collection of any organic compound or mixture of compounds. A 50 or 100 cm^3 gas tight syringe, fitted with a size No. 17 or 18 needle is conditioned to the atmosphere immediately before sampling by operating the plunger several times and the sample is then collected by inverting the syringe and allowing the plunger to fall under gravity over a period of about 3 min. The needle is sealed by insertion into a piece of silicone rubber and the sample returned to the laboratory for analysis. Several samples, for instance, as a survey or to check another method can be taken this way, sealed and stored for a short period. Samples collected thus should be analysed on the same day unless storage trials have shown that longer storage is possible.

Gas burettes

Gas burettes of volumes up to 500 cm^3 are normally made of glass with a tap at each end and sometimes fitted with a septum for sample withdrawal. Samples are collected by drawing the atmosphere through the burette using a hand or battery operated pump. To ensure a representative sample and to overcome adsorption a volume at least ten times that of the burette should be drawn through before closing the taps. As with the gas-tight syringe several samples can be taken and returned to the laboratory for analysis. Samples can be withdrawn via the septum for gas chromatographic or other analysis or reagents can be introduced for adsorption or to produce a colour reaction through the septum or one of the taps. Apart from possible surface adsorption of compounds on to the glass it is possible to lose sample by absorption or reaction with any greases used on the taps so that ungreased taps with PTFE keys are recommended.

A variation of the gas burette is to use a round bottomed flask fitted with PTFE keyed taps. A method is described for the determination of chlorinated aliphatic hydrocarbons such as trichloroethylene, carbon tetrachloride and perchloroethylene in the range 1–100 $\mu g\ cm^{-3}$ air using such a flask where it is claimed that the samples, if kept below 25 °C and stored in the dark, can be analysed up to three days after collection.[9] Samples are withdrawn using a 2 cm^3 gas syringe and analysed by gas chromatography.

Free diffusion bottle

A free diffusion bottle is a small (25 cm^3) glass container fitted with a screw cap which takes a PTFE backed septum. It can be used to take samples from inaccessible places and samples can be collected from several places simultaneously. The bottle is placed in the desired place (site) with the top removed and left for periods of 30 min or more. At the end of the sampling time the cap plus septum is replaced (the PTFE surface of the septum must be in contact with the sample) and

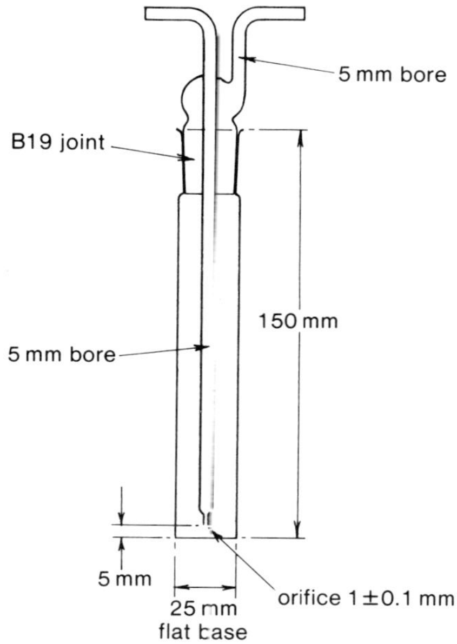

Fig. 4.2 Simple jet absorber (reproduced from *The Determination of Toxic Substances in Air: a Manual of ICI Practice,* with permission of W. Heffer and Sons Ltd.).

the sample returned to the laboratory for analysis. A sample of not more than 2 cm^3 is withdrawn with a gas-tight syringe and analysed by, for example, gas chromatography. This method provides a means for obtaining several samples at different locations simultaneously or of easily obtaining consecutive samples at a location. Samples should be analysed as soon as possible after collection and should not be left for more than 24 h.

Bubbler/impinger and reagent solution

There are so many possible combinations of bubblers plus reagents for determining toxic compounds in atmosphere[18] that it is practicable to give only some examples of applications. Before carrying out analysis, the user must decide on the best reagent for a particular application (bearing in mind what equipment is available for subsequent analysis). Then, knowing the compound's toxicity and the sensitivity of the method, the user must decide the flow-rate and time required. The flow-rate needed will decide the type of bubbler/impinger required and some typical designs are shown in Figs 4.2 and 4.3. The impinger shown in Fig. 4.2 is used to determine aldehydes and ketones in atmosphere the reagent being 2,4-dinitrophenylhydrazine and the hydrazone derivatives being determined using liquid chromatographic methods. A sample time of 10 min at a flow rate of 500 cm^3 min^{-1} gives typical limits of detection of 0.05 p.p.m. v/v for acrolein, acetaldehyde, formaldehyde.

The bubbler shown in Fig. 4.3 is used to determine mercury in atmosphere by absorption into acidic permanganate and subsequent analysis by dithiozone or flameless atomic adsorption.

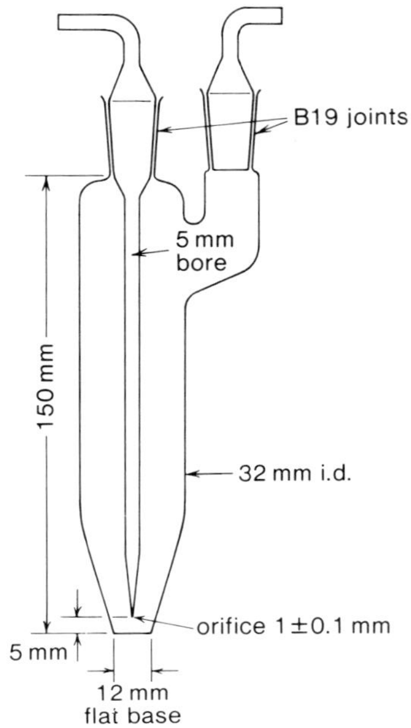

Fig. 4.3 Fast rate absorber (reproduced from *The Determination of Toxic Substances in Air: a Manual of ICI Practice,* with permission of W. Heffer and Sons Ltd.).

It will be obvious that in a bubbler the sampling of the contaminant vapour takes place by transfer of that vapour from the gas phase in the bubble to the liquid across the interface. Therefore, for efficient sampling, the following conditions are essential.

(1) The interfacial area between the gas and liquid should be as high as possible. This means that for a given flow rate, the bubbles must be as small as possible.

(2) The contact time between the gas and liquid should be as long as possible. This means the flow-rate through the liquid should be as low as possible.

(3) The partition coefficient of the substances between the gas and liquid phase should be such that substance has high solubility in the liquid phase and low partial pressure in the gas phase.

Requirement (1) can be achieved by breaking up the gas flow with a frit or using a small diameter orifice into the liquid. Requirement (2) can give problems in obtaining enough sample for subsequent analysis. This, of course, depends on the sensitivity of the analytical technique which is to be used and emphasises in this form of measurement technique, the interdependence of sampling and analytical techniques. Requirement (3) is often met by using a chemical reagent in the sampling liquid so that the contaminent vapour forms a reaction product in the liquid phase. Where the vapour has a significant partial pressure in the gas phase above the absorbent liquid, it may be necessary to use more than one sampler in series. Moreover, in such cases, the problem of sample loss will be increased if the vapour

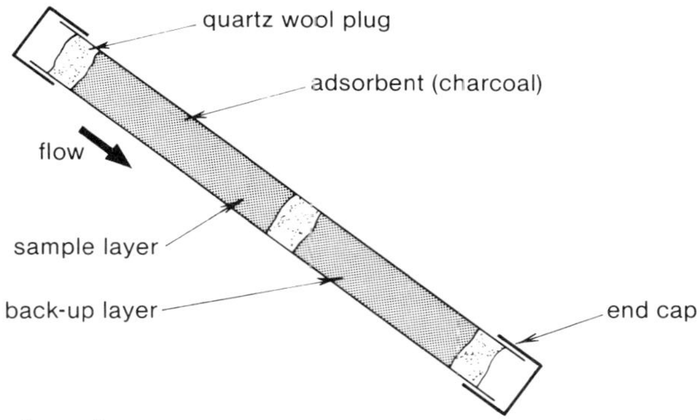

Fig. 4.4 Adsorption tube.

pressure of the absorbant is high enough for substantial evaporation of it to occur during sampling.

Solid adsorbents

Samples, either short term or long term, can be collected on adsorbents packed in a suitable tube[19,20] (Fig. 4.4) and this technique, although used for background measurements, has particular usefulness for personal monitoring and is described in more detail in Chapter 5.

There is a wide range of adsorbents, some common ones being carbon, Porapak, Tenax and silica gel. Tubes within the size range 2 mm i.d. × 4 cm to 4 mm i.d. × 15 cm (holds approximately 1 g carbon) are commonly used and can be made of a variety of non-reactive materials. The type of tube and adsorbent used will depend on the application and the method of analysis.

Some of the smaller tubes can be placed directly into a gas chromatograph linked to a mass spectrometer for identification and analysis.

Care should be taken in the choice of adsorbent with respect to: breakthrough volume (that is, how much sample can be drawn into the tube before the toxic gas or vapour passes through completely), humidity (some adsorbents, such as silica gel and carbon will adsorb water and become less active), and temperature. These and other parameters are discussed in detail in Chapter 5.

The methods used to remove the adsorbed gases from the tubes are:

(i) *Thermal desorption.* The tube plus adsorbent is heated to a sufficiently high temperature to remove the compounds of interest. The desorbed gases can, for example, be fed into a gas chromatograph, the cell of an infra-red analyser or a detector tube.[15]

(ii) *Solvent desorption.* The tube contents are transferred to a vial and a minimum volume of a suitable solvent such as carbon disulphide, methanol or acetone is added. The mixture is shaken and allowed to desorb for a period determined by experiment – 1 h is typical. Aliquots of the extraction solvent are then analysed using, for example, gas chromatography, infra-red or ultra-violet spectrophotometry.

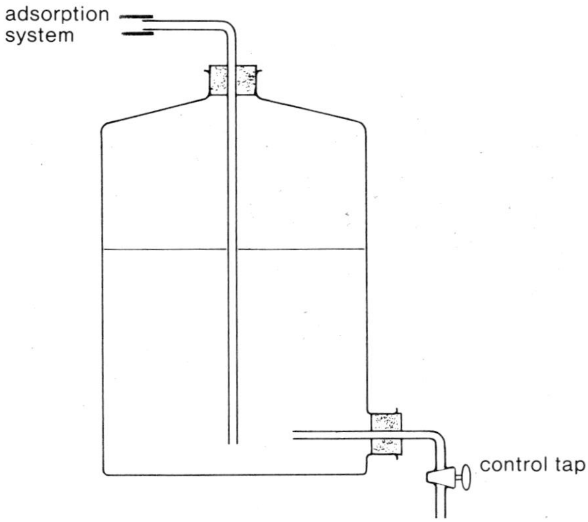

Fig. 4.5 Water aspirator.

Reaction tube

This is entirely analogous to the adsorption tube, except that the packing reacts chemically with the toxic vapour to form an involatile product. For example, acidified silica gel could be used to trap amines or basified silica gel to trap sulphur dioxide, chlorine, etc. With an aqueous-based reagent, the tube contents can usually be extracted with hot or cold water for analysis.

Water aspirator[2–4]

A water aspirator is cumbersome but useful for collecting large sample volumes (up to 10 dm^3) and for sampling at low rates, and is used to draw the atmosphere through a bubbler or adsorption tube. It can be used in places where no power supply is available to operate a pump system and can be used in an area where flammable gases could be present. An aspirator is shown in Fig. 4.5, the adsorption system is connected to the inlet tube which almost reaches the bottom of the aspirator to give a constant head of liquid. The tap at the bottom of the aspirator is opened and water allowed to run out at the desired rate, which will draw atmosphere into the adsorption system. The volume of water discharged from the aspirator is a measure of the volume of air sampled. The aspirator can be calibrated and marked or the volume discharged collected in a measuring cylinder.

Plastic bags

Samples can be collected in plastic bags[13] made from materials such as PVC (polyvinyl chloride), Tedlar (polyvinyl-fluoride), Saran (vinyl-vinylidine chloride co-polymer) and Mylar (polymer film). Bags vary from about 1 dm^3 up to about 50 dm^3 and sample time is at the discretion of the user but can range from seconds to hours, depending on the situation. When using a plastic bag the possibility must be considered of absorption into or permeation through the bag material by the toxic substance or reaction of the toxic substance with, for instance, atmospheric moisture must be considered. When small samples are taken the bags can be 'purged' with atmosphere which will help reduce adsorption problems. Equipment used to fill the bags with toxic gases such as pumps must also be regarded with care with respect to sample adsorption or reaction. Unless storage experiments are carried out, the time between sampling and analysis should be as short as possible. Plastic bags have advantages over other collection systems in being light, inexpensive and allow the sample to be withdrawn without dilution by replacement air or reduction of pressure. Sample loss by permeation can be eliminated by having layers of aluminium foil in the plastic.

3.2 Continuous or rapid repetitive monitors

3.2.1 Introduction

Preferably these monitors should be battery-operated to make them fully-portable, but it is feasible to use mains-powered instruments either keeping the instrument fixed and using a long, flexible sample line or carrying the instrument and using a long mains lead. Because there are no power limitations, reduction in sampling rate from long sample lines can be compensated by using more powerful pumps.

When using a continuous or rapid repetitive monitor to do hygiene measurements in the work-place the user must be aware of the following points.

(1) Is the area one where only intrinsically safe instruments can be used? If so then, either monitors with appropriate certification must be used or the instrument will have to be enclosed in a purged and interlocked box and samples taken remotely using long, flexible sample lines (beware of adsorption on sampling lines).

(2) Is the instrument mains or battery-operated? If mains operated ensure that the correct supply is available on plant (unless provided with suitable barriers, a mains-powered instrument cannot be used in flammable areas). Approved battery-operated equipment cannot be run off a charger in a flammable area.

3.2.2 Paper tape analysers

These devices work on the same basic principle as the test papers described on p. 93. Air containing the toxic substance is drawn through a slit and then through a moving paper tape impregnated with reagents to make it sensitive and selective to the substance of interest. The tape, which moves at a constant rate, develops a coloured stain of intensity related to the concentration of toxic substance in the air sampled. The monitor uses a double beam optical-measuring principle (Fig. 4.6) in which the top half of the tape is exposed to atmosphere while the bottom half remains unexposed. Two light beams from a single source are shone on to the tape and the reference and measurement photometric detectors in the optics block sense light reflected from the unexposed and exposed portions of the tape respectively and produce a differential electronic output signal which is proportional to stain intensity and, thus related to concentration of the toxic substance. The signal is indicated on a

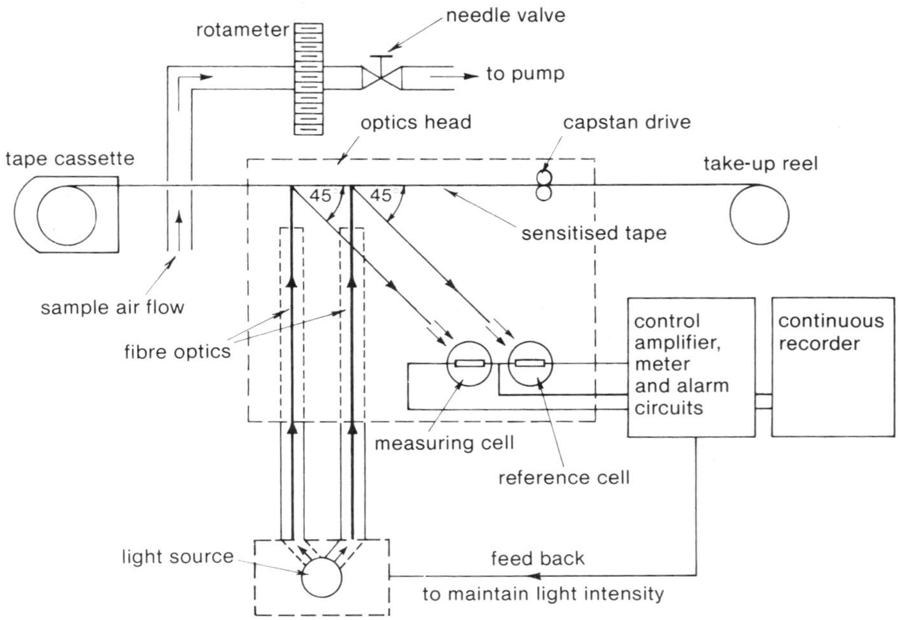

Fig. 4.6 Schematic diagram of paper-tape analyser (redrawn with permission of MDA Scientific (UK) Ltd.).

meter mounted on the front panel of the monitor, and recorded by a strip chart recorder to produce a permanent real time/concentration profile.

The monitors are fitted with self-contained audio and visual alarms which are activated if a preset concentration is exceeded. They are mains operated so are not intrinsically safe but can be installed in an enclosure system and used with a power unit. This type of instrument has the obvious advantage that by using a range of tapes the same instrument can be used to measure a number of gases. At present, tapes are available for monitoring phosgene, toluene di-isocyanate, hydrogen chloride, hydrogen sulphide, chlorine, sulphur dioxide, ammonia, hydrazine, toluene diamine, organic isocyanates (total), vinyl chloride, dinitrotoluene and nitrogen dioxide and the range is continuously being extended. The development of paper tapes is not simply the transferance of liquid-phase colour chemistry methods to paper or the transfer of colorimetric tube techniques. The main problem is in obtaining tapes with a suitable storage life – say six months before use. Paper itself has many functional groups such as hydroxyl on its surface, which will react with the reagents in addition to the reaction of the reagents with the cellulose itself. The reactions of the vapour with the reagents, generally require the presence of water and hence the addition of humectants such as glycerol to the paper and this water increases the reaction of the reagents with the paper. It may be that more inert support matrices such as woven silica, or even the adaptation of impregnated gel films along the lines of photographic films could extend the range of applicability of this technique.

Continuous tape monitors can be checked in two ways: (i) electronically using a test circuit which checks the optical system and electronic circuitry and indicates when adjustment is necessary; (ii) secondary calibration is accomplished using a standard stain on a calibration test strip. However, it is desirable that the complete system be calibrated by drawing a standard atmosphere (Chapter 7) of the toxic substance in via the sample inlet.

A major limitation of the paper tape analyser for conducting surveys or searching for leaks is the built-in time lag between sampling the gas and reading the intensity of the stain. This can be several minutes. This time lag is due to (*a*) the finite width of the sampling slit, (*b*) the slow movement of the tape and (*c*) the time taken for the tape to develop maximum colour intensity at the sampling rate of the atmosphere. Factor (*c*) is the critical one. If the flow rate is increased, then the time taken to develop a given intensity reduces, the tape speed can be increased and hence the response time reduced. However, the flow rate cannot be increased indefinitely owing to the flow resistance of the paper and slippage of the contaminant vapour through the paper without reaction. The most sure way of reducing response times is to develop paper tape or other systems where the development of the stain is much more rapid than in present types. In this way the sampling slit can be reduced in size, the tape speed increased with resultant decrease in response time.

3.2.3 Infra-red analysers

These analysers are described in detail in Chapter 2 and can be used to measure any gas which has absorption bands in the infra-red region of the spectrum.[21] Infra-red spectra of most compounds are complex but each spectrum contains only a small number of strong absorption bands which can be seen at the low concentrations encountered in ambient air analysis. Infra-red analysis always gives an indication at a characteristic specific wavelength for a given compound. Absorption bands can serve to positively identify the compound as well as indicate the amount present.

Single beam infra-red instruments are available using continuously variable interference filters (Chapter 2). When fitted with a variable path gas cell, these analysers are useful tools for continuous monitoring of toxic gas in atmospheres in the range sub-p.p.m. upwards.

The analysers give a direct-reading output of absorbance either on a built-in meter or to an external potentiometric recorder. They need to be calibrated for the gas of

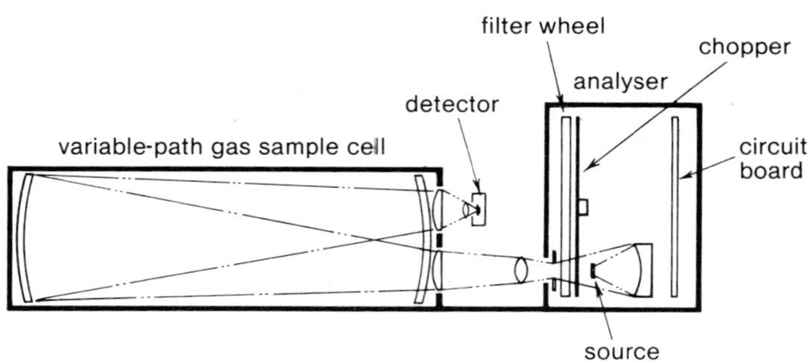

Fig. 4.7 Miran Gas Analyser (redrawn with permission of Foxboro/Wilks Inc.).

interest either using standard atmospheres, or by use of a closed calibration loop supplied as an accessory, to give a graph of absorbance versus concentration.

For field use, the analysers have a built-in pump which draws sample atmosphere continuously through the gas cell and the resultant meter or recorder reading is related to concentration using the calibration graph. It is possible to get dedicated instruments in which the meter is directly engraved with a concentration scale. The range of application of such instruments is illustrated in Table 4.1 using data for the Miran range of instruments marketed by Foxboro/Wilks (Fig. 4.7).

3.2.4 Ultra-violet analysers

In principle these are analogous to infra-red analysers, but, because of the lack of specificity of ultra-violet absorption compared with infra-red, they are less widely used. In fact, the only common use is for mercury analysis, utilising the very narrow band-width of mercury emission and absorption.

Table 4.1 Variants on the Miran analyser theme

Instrument parameter	*Model*				
	1A	*104*	*103*	*101*	*80*[5]
Wavelength range (μm)	2.5–14.5	2.5–14.5	—[2]	—[4]	2.5–14.5
Absorbance ranges	0–0.025 to 1–2	0–0.025 to 1–2	NA	NA	0–2
Transmission range (%)	0–100	0–100	NA	NA	0–100
Cell path length range (metres)	0.75–20.25	0.5–10.5	0.5–14.5[3]	0.5–14.5[3]	0.75–20.25
Power	Mains	Mains	Mains or Battery/ inverter	Rechargeable battery or mains	Mains
Weight (kg)	14	11	10	8	24
Cell volume (dm^3)	5.6	2.5	2.5	2.5	5.6
Recorder output for FSD (V)	0–1	0–1	Not specified	None	0–1[6]
Maximum sample line length (metres)	30	30	3	30	30
Concentration range	—[1]	—[1]	20–400 p.p.m. v/v FSD	$<$1 p.p.m. v/v upwards	—[1]
Operating pressure (kN m^{-2})	Vacuum to 1013	Vacuum to 133.3	Vacuum to 106.6	Vacuum to 106.6	Vacuum to 1013

[1] Concentration range dependent on vapour being analysed, typically sub-p.p.m. upwards.
[2] Pre-set by interchangeable selective filter.
[3] Adjusted in factory within this range.
[4] Adjusted in factory within range 2.5–14.5 μm.
[5] Microprocessor-controlled multi component analyser with correction for cross-interferences for up to eleven components.
[6] 0–1 Volt for 0–100% transmission range. To convert to absorbance a logarithmic converter is needed. The Optilab Multianalog 201 is suitable.

A typical analyser similar in principle to the infra-red analysers on p. 104, uses a mercury-vapour lamp to give 0.2537 μm radiation which is absorbed by mercury vapour in the sample drawn through a silicon cell and detected by a photo-resistive cell.

The photo-resistive element forms one arm of a Wheatstone bridge and when the element changes, because of absorption of radiation by sample, the bridge becomes unbalanced which causes a meter deflection. The meter is calibrated over the ranges 0–0.2 mg m^{-3} and 0–1.0 mg m^{-3} mercury and has a built in filter to facilitate zeroing in the field.

3.2.5 *Photoionisation analysers*

Photoionisation can be used to measure a wide range of toxic compounds. The process is called photoionisation because the absorption of ultra-violet light by a molecule leads to ionisation:

$$R + h\nu \rightarrow R^+ + e^-$$

where R^+ is the ionised species and $h\nu$ represents a photon which has an energy greater than the ionisation potential of the compound. The sensor is a sealed ultra-violet light source that emits photons with sufficient energy to ionise many compounds (particularly organic) but do not ionise the major components of air such as oxygen, nitrogen, carbon monoxide, carbon dioxide or water vapour. Generally the 10.2 eV emission line of hydrogen is used. A chamber adjacent to the ultra-violet source contains a pair of electrodes and when a positive potential is applied to one electrode the field created drives any ions formed by absorption of the ultra-violet light to the collector electrode, and the current which is proportional to concentration is measured. A typical photoioniser operating on this principle weighs about 4 kg, with a detection limit of about 0.1 p.p.m. v/v, and a useful range up to 2000 p.p.m. v/v. Concentration is read directly on a linear meter on the instrument or can be fed to a recorder. Although not intrinsically safe the system has the advantage over flame ionisation detectors described below in that it does not respond to methane which is always present in ambient air, and has some degree of specificity in that the energy of the photons can in principle be chosen to correspond with the ionisation potential of a given material by using a suitable emission source. In addition, the device does not require a supply of gas.

3.2.6 *Electrochemical analysers*

Electrochemical analysers are well known in applications for the determination of oxygen in atmosphere (Chapter 4). Detectors have been developed[12] for use in portable and personal instruments to measure carbon monoxide in air in the p.p.m. range. Air containing carbon monoxide is drawn through a detector cell containing a catalytic anode where it is oxidised to carbon dioxide.

$$CO + H_2O = CO_2 + 2H^+ + 2e^-$$

The current resulting from the oxidation is directly proportional to the CO concentration which can be read directly off the instrument meter (Fig. 4.8) and a detection limit of about 1 p.p.m. of CO in air can be achieved.

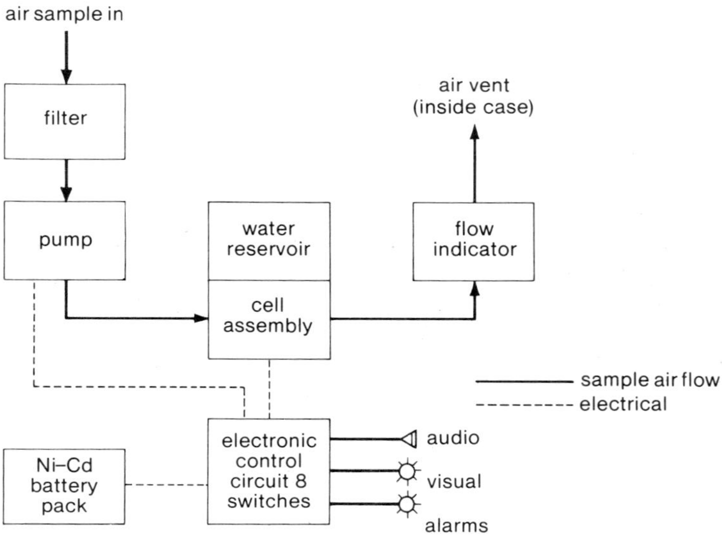

Fig. 4.8 Carbon monoxide dosimeter functional schematic (diagram redrawn with permission of MDA Scientific (UK) Ltd.).

Other portable instruments are available for the electrochemical determination of chlorine, hydrogen sulphide, hydrogen cyanide, carbon monoxide and nitrogen dioxide and are described in more detail in Chapter 5.

3.2.7 Flame ionisation detectors (FIDs)

Gas chromatographic units fitted with FIDs are extremely useful for continuous or rapid repetitive measurements of organic (toxic) compounds in ambient air. Where it is not necessary to separate the different components of the atmosphere before measurement either because only one main contaminant is present or because it is sufficient to measure the total concentration of organic contaminants, then the FID can be used by itself for measurements.

As its name implies, the FID involves the ionisation of organic components of the atmosphere in a flame. A hydrogen/air flame is used because the concentration of ions within it is low. The introduction of organic carbon compounds into the flame results in the formation of ions which reduces the electrical resistance of the flame by many orders of magnitude. This change in resistance is measured using two electrodes adjacent to the flame. The ionisation processes in the flame are complex but only about 10^{-6} v/v of the organic material burnt appears as ions. The formation of such ions has been found to be linearly related to the concentration of organic material as long as this does not exceed about 1% of the flammable gas supply to the flame.

A typical FID is a small chamber with a hydrogen flame on a burner tip. A collector electrode which is usually the anode is positioned above and around the flame and the burner tip is usually the cathode. The flame conductivity is related by calibration to the concentration of carbon species in the atmosphere entering the chamber. The sensitivity of modern electronics is such that a pulse caused by 10 ng of carbon-containing compound can be measured.

This device will give a continuous measure of the atmospheric gas concentration and is the basis of many commercially available portable instruments for use in the field.

3.2.8 Electron capture detector (ECD)

An ECD contains a radioactive source, usually ^{3}H or ^{63}Ni which emits electrons. The cloud of electrons produced in the detector cell gives a current (called the 'standing current') which is shown on a built-in meter when a voltage is applied to the cell. When an electron-capturing material, generally halogen-containing compounds, enters the cell, electrons are absorbed and the standing current falls in proportion to the concentration and capturing efficiency of electron-capturing material. The detector can be calibrated using standard atmospheres of the material to be monitored and has a sensitivity of about 10 pg.

ECDs can be used for leak seeking or, after calibration, for quantitative measurement.[21] Instruments can be used for continuous monitoring when total electron-capturing material present in the atmosphere will be indicated. They can also be used with a sample valve and gas chromatographic column (at room temperature) to separate the electron-capturing materials from each other and from oxygen. Response is rapid; about 1 s for continuous monitoring and from 15 s upwards when used with a column.

Toxic materials which exhibit electron-capturing properties are usually halogenated (Arctons, carbon tetrachloride, trichloroethylene) so the nature of the detection system helps to make it specific. Care should be taken in use not to expose the detector to an 'overdose' as it will become saturated and take some time to recover.

3.2.9 Solid state detectors

The effects produced by the adsorption and reaction of gases on the surface of solid materials has given rise to two classes of sensors for gas detection, which are described in detail in Chapter 2. One type measures the heat of reaction produced by the oxidation of a gas on the surface of a catalytic solid (pellistor) and the second measures the change in electrical conductivity of a semi-conductor material produced by the trapping or release of charge carriers at the surface (normally incorporated in a Wheatstone bridge-type circuit for ease of measurement).

Pellistor detectors were originally developed for the measurement of exposive levels of gases and have limited applications for the detection of toxic compounds because of limitations in their sensitivity.

3.2.10 Semiconductors

A semiconductor can operate in two ways: (*a*) a reducing gas can be detected by a change in conductivity of the semiconductor associated with the removal of adsorbed oxygen reacting with the adsorbed gas and releasing electrons back to the solid which is detected as a conductivity change; (*b*) gases acting as electron donors such as water, hydrogen sulphide and sulphur dioxide adsorb on to the surface of the semiconductor and charge carriers are injected or trapped by the adsorbed gas.

Semiconductors are made from n-type oxides of the transition and heavy metals, they are normally metal doped or mixed with other oxides to improve selectivity and stability.

Instruments using semiconductors are available to measure a wide range of toxic compounds with sensitivities from 1 p.p.m. upwards. The signal from the detector is fed to a meter on the instrument which is calibrated in p.p.m. and can be multirange.

Problems with semiconductor detectors include the effects of humidity changes and a lack of selectivity in detectors developed to date.

4 Fixed-station monitoring

4.1 Introduction

This part of the chapter is different from the other 'applications parts' in that any fixed-point monitoring system has to be tailored to the measurement geography. Hence it is of little use to describe in detail any particular systems. Thus, it is best to concentrate on the philosophy and general approach to the concept of fixed-station monitoring and note that most of the measuring techniques described earlier can be used as the basis for such a system.

With the exception of small areas, fixed-station monitoring implies multipoint monitoring to give an adequate coverage of the area to be monitored. From this, two consequences emerge which must be taken into account *before* embarking on fixed-station monitoring: (*a*) the installation is likely to be expensive; (*b*) it will generate a lot of data. A monitoring system giving an analysis every 5 min generates over 100 000 data points per year.

From (*a*) it follows that management must be sure of the purpose for which fixed-site monitoring is to be installed. There are, broadly, two basic reasons for installing fixed-site monitors: to give early warning of high concentrations of a toxic vapour arising, for example, from unsuspected leaks; or to provide data from which exposure of personnel to toxic vapour can be assessed. For the former reason, the sampling sites should be located near potential sources of vapour, *whether or not people normally work in this area.* For the latter reason, sampling sites should be located in areas of highest occupancy of the working area. Some sites may, of course, serve both purposes.

From (*b*) it follows that some form of data-handling and reduction system is necessary, particularly for those monitoring sites to be used for calculating personnel exposure. For example, the data may be presented as averaged values together with records of highest exposure and/or number of times a pre-determined exposure level is reached or exceeded and these considerations are dealt with in more detail in Chapter 6.

4.2 How to achieve multipoint monitoring

There are three ways of achieving multipoint monitoring. The first way is to have an analytical instrument at each chosen site and to relay its read-out, either electrically or pneumatically to a centre for data handling/processing. This method is only feasible if the instruments are relatively cheap and may, of course, require intrinsically-safe instruments or instruments installed inside purged and inter-locked cabinets. Unless stability of zero and calibration are of a very high order, means must be provided for correcting for drifts in both. This implies setting up a system at each monitoring site which allows clean air and a standard atmosphere to be fed to the instrument's inlet and for adjusting both zero and sensitivity accordingly. This is a major disadvantage of the multi-instrument approach.

The second way is to have a centrally-situated instrument/data handling system and to pump atmosphere samples sequentially to this instrument from the fixed sampling sites. This approach is the only feasible one for expensive analytical instruments. It has the advantage that, since the instrument can usually be located in a non-flammable area, it need not be intrinsically safe, though it may be necessary to install flame-traps in the sample lines. Its big disadvantages are (i) the delay arising between sampling and analysis when long sample lines are needed even when large powerful sampling pumps are used; and (ii) the possibility of 'memory' effects arising from adsorption/desorption cycles in the sample lines.

A major advantage of the single-instrument installation is the ease of correcting for zero and/or calibration changes. All that is needed are two extra lines in the sequential sampling, one fed with clean air and one with a standard atmosphere together with automatic zero and sensitivity controls (which can be incorporated in the software of the data-handling system).

The third way is really a combination of the first two ways. This is the use of remote sensors, where available, at the sampling sites where output, normally electrical, is fed back to a central instrument/data-processing system. Here the instrument need not be intrinsically safe but, for sample sites in flammable areas, the sensors will need to be intrinsically safe and signal cables will need to be fitted with appropriate 'barriers'. This approach combines the advantages of the other two and overcomes most of their disadvantages except for the difficulty, described above, of correcting for zero and calibration shifts which are common with the multi-instrument approach. At present, the range of suitable sensors of adequate selectivity is very small. Future developments in analytical uses of semiconductors are the most hopeful way of improving this situation.

4.3 Choice of instrument

For the multi-instrument approach, choice is limited mainly by price to such instruments as moving paper-tape analysers, galvanic or polarographic instruments, katharometers, etc.

For the single-instrument approach, a much wider choice of instrument is available; gas chromatography, infra-red or ultra-violet spectrometry, mass spectrometry, etc. This wide choice of instruments gives increased selectivity together with the possibility of multi-component analysis. A number of instrument manufacturers offer instruments/packages for multipoint area monitoring using gas chromatography, infra-red spectrometry and mass spectrometry.

4.4 Area monitoring without sample sites

There is considerable development work going on currently on the concept of area monitoring by the use of open-path infra-red spectrometry using laser beams (see Chapter 3). At the time of writing, the authors do not know of any commercially-available package for this type of monitoring. It offers advantages of simplicity in sample-handling but suffers disadvantages in interpretation of results on a geographical basis.

5 References

1. BS 5343:1976. *Gas detector tubes.* British Standards Institution, London.
2. *Air Sampling Instruments for Evaluation of Atmospheric Contaminants,* 4th edn. American Conference of Governmental Industrial Hygienists, Cincinnati, Ohio (1972).
3. *Methods of Air Sampling and Analysis,* 2nd edn., ed. M. Katz. APHA Intersociety Committee. American Public Health Association, Washington DC (1977).
4. *The Determination of Toxic Substances in Air: a Manual of ICI Practice*, ed. N. W. Hanson, D. A. Reilly and H. E. Stagg. Heffer, Cambridge (1965).
5. A. H. Principe. The Vacu-Sampler, A New Device for the Encapsulation of Breath and Other Gaseous Samples. Presented at the 25th Annual Meeting of the American Academy of Forensic Science, Las Vegas, Nevada, Feb. 1973.
6. R. W. Miller *et al.* Vacu-Samplers, a Simple Method for Taking Accurate Quantitative Gas and Vapour Samples Using Partially Evacuated Steel Containers. Presented at the American Industrial Hygiene Conference, Minneapolis, Mn., June 1975.
7. F. W. Williams, J. P. Stone and H. G. Eaton. Personal Atmospheric Gas Sampler Using the Critical Orifice Concept. *Analyt. Chem.* **48**(2) (1976).
8. *The Determination of Vinyl Chloride: a Plant Manual*, ed. W. Thain. Chemical Industries Association, London SE1.
9. The Bureau International Technique des Solvants Chlores, B.I.T.-S.C. (Belgium). Standardisation of methods for the determination of traces of some volatile chlorinated aliphatic hydrocarbons in air and water by gas chromatography. *Analyt. Chim. Acta,* **82**, 1–17 (1976).
10. *Methods for the Detection of Toxic Substances in Air: Booklet No. 1: hydrogen sulphide.* HMSO (0000).
11. R. Hill, J. H. Holt and B. Miller. Rapid test paper methods for hydrogen cyanide in the atmosphere. *Annals Occup. Hyg.* **14**, 289–94 (1971).
12. H. W. Bay, K. F. Blurton, H. C. Lieb and H. G. Oswin. Electrochemical measurement of carbon monoxide. *Amer. Lab.,* July 1972.
13. F. J. Schuette. Plastic bags for collection of gas samples. *Atmos. Environ.* **1**, 515 (1967).
14. K. Leichnitz. Use of detector tubes under extreme conditions (humidity, pressure, temperature). *Amer. Ind. Hyg. Ass. J.* **38**, December 1977.
15. Personal communication.
16. K. Leichnitz. *Detector Tube Handbook, Air Investigations and Technical Gas Analysis with Drager Tubes*, 3rd edn. (1976).
17. *Gastec Precision Gas Detector System.* 'Blue Book' Gastec (ex D. A. Pitman Ltd., Agents) (0000).
18. *NIOSH Manual of Analytical Methods*, 2nd edn., vols. 1–4, ed. D. G. Taylor. US Dept of Health Education and Welfare, Public Health Service, NIOSH, Cincinnati, Ohio (1977).
19. A. Gold, C. E. Dube and R. B. Perri. Solid sorbent for sampling acrolein in air. *Analyt. Chem.* **50**, 1839 (1978).
20. E. D. Pellizzari *et al. Analyt. Lett.* **9**, 45 (1976).
21. G. G. Esposito *et al. Analyt. Chem.* **49**, 1774 (1977).

22. S. J. Silk, E. C. Hunt, A. F. Smith, R. Wood, F. A. P. Maggs and R. J. Sherwood. *Annals Occup. Hyg.* **16**, 51 (1973).
23. H. L. Kusnetz, B. E. Saltzman and M. E. Lanier. *Amer. Ind. Hyg. Ass. J.* **21**, 36 (1960).
24. B. E. Saltzman. *Amer. Ind. Hyg. Ass. J.* **23**, 112 (1962).
25. B. E. Saltzman. In *Air Sampling Instruments,* 4th edn. ACGIH (1972).

5 Personal monitoring

D. T. Coker

1 Introduction

1.1 General background

Nearly all developed countries have statutory Acts covering the health and safety of employed persons at work and the general philosophy behind these Acts is broadly similar in different countries.

One of the main responsibilities of employers arising from such Acts is to ensure that the air at the workplace is uncontaminated and fit for breathing, judged by local air quality standards, as discussed in Chapter 1.

There is unfortunately no generally accepted standardised procedure for air quality monitoring, so that the ways in which each employer sets about fulfilling his responsibilities are very diverse and are typically based on several factors.

(1) Different industries will naturally dictate different approaches and degrees of effort. Thus, a chemical manufacturing industry making or handling toxic volatile materials will obviously require more effort than a plant assembling mechanical components.

(2) The management's or health professional's interpretation of responsibility will probably differ from company to company.

(3) The resources and expertise which a company has available or can afford will vary widely as air quality monitoring requires manpower, capital equipment, laboratory facilities and a reasonably high level of expertise.

Setting up a monitoring programme to ensure that the air which employees breathe meets the required standards necessitates sampling being carried out on the person, as close to the nose and mouth as possible so that the level of contaminant substances measured in the sample closely corresponds to the levels inhaled by the person. This is often a contentious issue as it makes employees aware that they may be exposed to contaminant substances. This in turn means that they will want to know their exposure levels. Wearing sampling pumps and samplers can also cause problems of extra weight and hindrance, particularly on a person who may already have to put up with bulky protective clothing, ear muffs and hard hats, etc. There is a greater temptation to tamper with personal samplers and to load artificially larger amounts of contaminant. Tampering is usually overdone, however, resulting in obviously over-size results.

In spite of these problems, it is clear that personal monitoring is the most satisfactory way of measuring personal exposure.[1–3] Sampling air at fixed locations around the workplace is generally not representative of the air inhaled by individuals except in situations where the level of a contaminant is very uniform throughout the workplace or where people do not move around, but the relationship between measurements made at single points in space and time and personal exposure are described in Chapter 6.

These problems can be properly resolved only by good communications between management, workforce and those responsible for the monitoring. It is essential that employees are kept well informed and the joint works' councils are involved in all aspects of the monitoring, including the extent of the workplace problems, the rationale behind the monitoring programme, the results of the monitoring and the actions arising from the survey results.

This chapter will try to rationalise the type of monitoring approach for a particular type of hazard, describe the currently available techniques for sampling, analysis and measurement and indicate the direction of future techniques.

1.2 Monitoring programmes

Before any attempt is made to set up a monitoring programme, it is necessary to understand clearly the reasons for monitoring, the objectives of the programme and the purposes for which the resulting data will be used. It is easy to fall into the trap of the 'generating numbers' game and to lose the perspective view of what is being realistically achieved in overcoming the problems and hazards of the particular workplace.

There are several basic identifiable reasons for air quality monitoring:

(*a*) To protect the worker from sickness, impaired health, discomfort and inefficiency due to working conditions;

(*b*) to fulfil the Company's responsibilities under the local Health and Safety legislation;

(*c*) to ensure compliance with air quality standards;

(*d*) to establish an ongoing record of employee exposure levels;

(*e*) to indicate areas and operations where respiratory protection may be necessary;

(*f*) to provide data for epidemiological surveys, where medical records are analysed with exposure data. The ultimate test for assessing the correct setting of exposure standards is the correlation of measured levels of exposure with the occurrence or absence of clinical symptoms. At present few air quality standards have a sound basis in human experience.

The scope of a personal monitoring programme should be to define the mean exposure and the range of exposures for each job situation under all ranges of plant operating conditions and weather conditions.

Survey results should be used as a guide for improving conditions where results are unacceptably high. The specific operations and processes leading to high exposures can be traced by making measurements during each different work task and coupled with studies of working practices. This can often point the way towards re-designing these practices in order to minimise exposure.

1.3 Monitoring techniques

The prime consideration in deciding on the type of monitoring technique to be used is the nature of the toxic hazard from the contaminant vapours likely to be present in the workplace air. These hazards can be broadly split into two classes.

(*a*) *Acute toxic hazards*, where overexposure results in the rapid occurrence of toxic symptoms. A typical example is H_2S which causes irritation and dizziness within a few minutes at levels of a few hundred p.p.m. and rapid loss of consciousness followed by death within a few minutes at levels around 1000 p.p.m.

(*b*) *Chronic toxic hazards*, where prolonged or repeated overexposure can result in the onset of toxic symptoms but where a single day's overexposure is unlikely to elicit a toxic response. A typical example is benzene where levels of a few hundred p.p.m. can be tolerated without any immediate effect but prolonged and repeated exposure to much lower levels can eventually lead to blood disorders and, in severe cases, leukaemia, a process which can take many years.

It is therefore evident that in order to protect a worker from an acute toxic hazard, such as H_2S, there is little point in monitoring the time-weighted average exposure where results may be obtained only days after the monitored period, whereas the risk to health from overexposure is immediate. The most appropriate type of monitor in this instance would be a fast-response audible/visual personal warning device set to trigger at a predetermined level.

Conversely this latter type of device may be unsuitable for monitoring a substance with chronic toxic effects as high-level transient exposure peaks may be occurring and repeatedly trigger the alarm, but the time-weighted average may be below the threshold limit value (TLV). In this case personal 'whole-shift' monitoring is the preferred technique, so that a check can be kept on the long-term cumulative dose.

In general personal monitoring techniques can be categorised into five types:

(1) *Rapid-response hazardous level warning devices.* These should be used to protect workers against substances which are immediately hazardous to health and where the toxic effect depends more on the actual concentration in the inhaled air rather than on the dose received by the body, e.g. H_2S, HCN, phosgene, high-level flammable gases, and oxygen deficiency.

(2) *Exposure monitors which measure an integrated concentration/time level.* Ideally these should sound an alarm when a particular concentration/time level is reached. They should be used for substances which are not likely to be immediately hazardous at normally encountered levels but which can rapidly build up in the body

to give toxic symptoms within periods of less than a workshift. Some types are however also used for immediately hazardous substances, but without a warning fitted they can only monitor exposure and do not warn the person of overexposure. Typical examples of this type are dosemeters for CO and NO. Another example is the caged canary in coal mines – probably the earliest example of a toxic gas warning system.

(3) *Direct-reading chemical stain devices.* These include long-term detector tubes, paper tape monitors and stain badges, where the length of stain along a tube or the intensity of stain on the paper or badge gives a rough measure of the accumulated dose/time level. A limited amount of protection is afforded by these devices as the person can check the reading from time to time. They can be used for monitoring many types of substances but are best used for the acutely toxic gases such as NH_3, CO, H_2S, HCN, etc.

(4) *Time weighted average (TWA) samplers.* These can sample the air continuously over long-term periods giving a sample representing the average level over the sampled period. The samples then require subsequent analysis. These can be used for monitoring most substances but are more commonly used for organic gases and vapours. They give protection only against chronic toxic effects by showing up previous overexposures which must be rectified.

(5) *Analysis of body fluids for personal monitoring.* In these the body is used as the sampler and the resulting concentrations of the contaminant or its metabolites in urine, blood, breath, etc. are used to estimate the level of exposure.

2 Safety considerations

For obvious reasons any equipment worn by a person must be completely safe and not add to any hazards in the workplace. These requirements can cover a number of safety aspects.

(*a*) *Electrical safety* is a general requirement in most industries where flammable materials are handled or where flammable gases or vapours can be present for any reason. The basic need is that *the equipment must not be a potential source of ignition* in its ability to cause an electrical spark of sufficient size and intensity to ignite a flammable vapour (incendive spark). This can be prevented ideally by making the equipment intrinsically safe so that its design makes it impossible in normal operation to release an incendive spark. The next best alternative is to use a flame-proof or explosion-proof enclosure through which a flame cannot propagate.

(*b*) *Mechanical safety* involves, for example, construction of the equipment so that it is not capable of generating an incendive spark when dropped on to a hard floor from its normal operating height and will also not break open to allow an electrical spark to occur.

(*c*) *Functional safety* is another requirement. A warning device should always 'failsafe', i.e. in the event of its failing it should indicate its failure, usually by sounding-off its alarm, so that the wearer is warned that it is no longer protecting him. Similarly sampling pumps and dosemeters should be designed so that they indicate any flow interruptions or failures, in order to avoid false low results being reported.

Apart from these considerations the equipment should not have sharp edges or protrusions which could injure the wearer in the event of a fall or blow.

In designated areas, Intrinsic Safety Certification is a legal requirement and in the past many countries have had their own, different specifications. Fortunately the degree of standardisation is improving and the common EEC[4] Standard has now been accepted by most European countries so that, for instance, the UK BASEEFA[5] certificate which complies with the EEC Standard will be acceptable in most other EEC countries. The US MSHA (Mining Safety and Health) and FM (Factory Mutual Research Corporation) certification do not comply with the EEC Standard.

3 Personal monitoring techniques

3.1 Hazardous level warning meters and dosemeters

These are portable devices, worn in the top pocket or on a belt, and are used to warn the wearer of overexposure to substances which may be hazardous to health within a short time. They contain a detector cell which responds to the substance in proportion to its concentration, and circuitry which will trigger when a predetermined level is reached – usually the TLV or ceiling value – sounding off an audible alarm and/or a flashing light to warn the wearer of the hazardous level. There are three types of detector cell commonly used in these meters.

3.1.1 Filament types

These use either a plain catalytic filament or a metal oxide-coated filament (pellistor[7]) which responds to concentrations of any flammable vapour in air up to the lower explosive limit (LEL). They are discussed in detail in Chapter 2 and are typically used to protect the wearer from the flammability hazard of common relatively non-toxic fuel gases and vapours (natural gas, LPG, etc.), usually in enclosed situations such as empty storage or cargo tanks, underground pits and mines, etc. Their limits of detection are generally well above most existing TLVs and they should not therefore be used for protection against toxic vapours. The principle of the detector[64] is that, although combustion will not propagate through a mixture of air and flammable vapour at levels below the LEL, it can occur to a limited extent on the surface of a heated catalytic filament. This energy of combustion raises the temperature of the filament and also its resistance. The surface-catalysed combustion becomes more exothermic as the vapour concentration increases towards the LEL, at which point all the gas in the cell chamber might ignite. For this reason these detectors are not used in an open situation but are placed inside chambers protected by flame traps – generally sintered metal plugs – to prevent them causing an explosion.

The increase in cell temperature and resistance is proportional to vapour concentration and so the cell resistance is used as the output signal.

The advantages of this type of detector are its simplicity and its general applicability for flammable hazard situations. Its disadvantages are:

(*a*) lack of sensitivity;

(*b*) lack of specificity;

(*c*) failure to function properly in oxygen-deficient situations;

(*d*) sensitivity can be reduced by poisoning giving a loss of response (this is less of a problem with pellistors);

(*e*) high current drain, as the filament temperature needs to be maintained at several hundred degrees centigrade;

(*f*) unsatisfactory stability necessitating frequent calibration;

(*g*) wide variation of response to different vapours, depending on the type of functional groups, the LEL and the molecular weight. Within the members of a homologous series the response approximates most nearly to weight percentage of the vapour.

A variant of this type of cell is the Hopcalite catalyst which is specific for CO at low temperatures and gives almost no response to other flammable vapours.

Another less widely used filament detector is the long established thermal loss type[6] which does not operate on the catalytic combustion principle but relies on differences in thermal conductivity and convection between air and the contaminant vapour. These alter the rate of heat loss from a heated inert filament and give a resistance change similar to the explosimeter type. They are mostly useful for fairly high levels of non-toxic gases such as light hydrocarbons and CO_2 in the brewing industry.

3.1.2 Electrochemical detectors (see also Chapter 3)

These are based on the electrochemical cell which at its simplest consists of two electrodes immersed in an electrolyte. This type includes detector cells of a wide variety and subtype, e.g.

(*a*) *Conductivity cells.* These measure the increase in conductivity of a liquid when an ionisable substance is dissolved in it.

(*b*) *Coulometric cells.*[8] Here a gas reacts completely within a cell generating a current proportional to the amount dissolved, according to Faraday's Law. In this case measurements are made of the number of coulombs of current generated.

(*c*) *Galvanic cells.*[9] These operate in the reverse way to electrolysis, e.g. any oxygen applied to the cathode will generate a current.

(*d*) *Polarographic cells.*[10] These are based on oxidation-reduction reactions.

(*e*) *Fuel cells or diaphragm cells.*

(*f*) *Specific ion electrodes.*

These latter types overlap considerably and it is difficult to categorise each separately and unambiguously; they are really just terms which are used in practice and are not definitive in this context.

Conductivity cells are seldom used for personal monitors as they require bubblers and are used mainly for general acid gas determination.

Galvanic, polarographic and fuel cells are used almost exclusively for oxygen determination and form the basis of many personal oxygen deficiency meters. Oxygen deficiency is discussed in depth in Chapter 3. Briefly, oxygen levels below the normal (20.8% by volume) present a hazard to humans and it is generally accepted that 18–18.5% is the minimum allowable level for normal working. Below this symptoms will appear depending on the person's work rate and the level of O_2, starting with shortness of breath and impaired judgment and eventually coma and death at lower levels. In enclosed, shut-off spaces, oxygen can be depleted in many ways, e.g. by reaction with micro-organisms, corroding metals, oxidisable deposits (sulphides, etc.), preferential solution, etc. (oxygen is more soluble than nitrogen in water and most organic liquids).

These meters warn the wearer immediately if the oxygen level falls below an acceptable value. They also have applications in the rarer situations where an excess of oxygen may be present, giving an increased fire hazard and, at higher levels, a risk to health.

Coulometric cells cover a wide range of designs that would fill a whole book if they were treated individually. Their most common application is for monitoring inorganic gases such as CO, SO_2, NO, NO_2, NH_3, H_2S, Cl_2, $COCl_2$, O_3 and HCN where the toxic hazard is mainly of an acute nature. Many of these gases are irritants and have ceiling TLVs assigned which should not be exceeded even instantaneously. Coulometric cells are well suited to this application as they have a rapid response and can be made fairly selective.

Cells used in personal monitors do not usually have an external electrolyte system. The cell is normally sealed and contains the electrodes and electrolyte which can be liquid, in gel form or even solid. The gas to be monitored has to pass through a membrane into the cell.

Selectivity can be achieved in several ways: (*a*) by the choice of specific cell reactions; (*b*) by setting the electrode potential so that other gases with higher electrode potentials do not react; (*c*) by using a selective membrane which is permeable only to the particular substance.

An example of a warning device is the Compur Monitox 4100 for H_2S, which responds within a few seconds, and sounds an audible alarm at 10 p.p.m. The cell reaction is:

$$H_2S + 2\,Ag^+ \rightarrow Ag_2S + 2\,H^+$$

Both electrodes are silver, one being a diffusion electrode through which H_2S passes to react with the electrolyte, generating a current.

An example of the dosemeter type is the Ecolyser CO dosemeter where CO is electrochemically oxidised in a cell with concentrated sulphuric acid electrolyte and teflon-bonded platinum diffusion electrodes, plus a third reference electrode of mercury/mercury sulphate.

3.1.3 Solid state cells

These cells were evolved from the explosimeter detector where the platinum filament type was improved by coating it with a metal oxide to improve stability, sensitivity and resistance to poisoning. The next step was the elimination of the platinum filament which was replaced by a heating element and electrodes embedded in the bead. It was found that the resistance of the metal oxide bead itself changed when exposed to vapours, independently of the temperature effect, and the resulting sensors, made of oxides such as tin and zinc, are maintained at a steady temperature and used for detecting conbustible gases. They also respond to fine particulates such as soot and smoke particles and have thus been used as smoke and fire detectors. They can be made with a limited degree of selectivity by varying the composition of the bead and its operating temperature. Their mode of operation is described in Chapter 2.

The latest generation of solid state detectors has now evolved from an electrochemical principle. According to this, the heated oxide, usually of a transition metal in Group III and IV of the Periodic Table, acts as an electrolyte, and gas molecules diffusing into the cell are dissociated into ions or ion complexes producing a current through the cell between polarised electrodes.

These sensors are considerably more sensitive than the catalytic sensors, typically by two orders of magnitude. Their dynamic range is greater and oxygen is not necessary for their operation and so they are not limited to the LEL ceiling. This feature also eliminates cell burn-out as there is little temperature variation of the cell with gas concentration. Burn-out can be a problem with catalytic sensors if very high concentrations of flammable gas reach the cell. Sensor stability and resistance to poisoning are good as the cells operate at a lower temperature – typically 300–400 °C – compared with around 800 °C for catalytic sensors. They do not rely on catalytic activity so that they are less susceptible to poisoning or deterioration. Disadvantages of these detectors are slow response, long meter recovery time, typically up to three minutes, and a lengthy warm-up period, up to 30 minutes.

Example: IST H_2S meter (manufacturers data)
Minimum detectable concentration – 1 p.p.m.
Response time – 40 seconds to 90% of final reading.
Recovery time – 1 minute to 90%.
Main interference – methyl mercaptan.

Specificity is improved by selection of the appropriate metal oxide combination, the composition of the cell coating layer, the cell operating conditions and the use of external filters. Interference can occur: (*a*) in direct proportion to the concentration of the interfering substance – but usually at a lower indicated concentration; (*b*) when the concentration of the interfering substance exceeds a ‘no effect’ threshold level. In general specificity is best for the permanent gases and worst for a specific organic compound in the presence of its homologues.

3.1.4 ‘Semiconductor’ detector

This is a recent development[12] of the solid state detector which operates on the same ionisation principle as the metal oxide bead cells, but uses a silicon semiconductor base containing two electrodes, on to which a thin layer of transition metal oxide is vacuum-deposited. The operating temperature is below 100 °C. This sensor has a similar sensitivity to the metal oxide bead type but is claimed to have a much more rapid response (around 1 s) and a much higher selectivity for specific compounds.

3.2 Chemical stain devices

These are based on a chemical reagent system adsorbed on to an inert support, such as silica gel, alumina or paper. The reagent system is designed ideally to react only with the substance to be monitored. Thereupon a stain is developed, the size or intensity of which is proportional to the level of the substance in the air during the sampling period.

At their simplest, these devices consist of a badge containing a piece of filter paper impregnated with a chemical, the most commonly used being lead acetate to detect H_2S, which gives a brown stain. The intensity of the stain is a rough measure of the average H_2S level in the air, which can be measured by comparing the stain with a calibration chart. Transport of H_2S on to the badge is by diffusion, but the stain development rate is controlled largely by the reaction kinetics. These are called 'passive' samplers because air is not actively driven on to the sampling medium. A refinement of this type is a badge where the stain spreads along a strip of paper overprinted with calibration marks.

3.2.1 Detector tubes[66,67]

The most common type of stain device is the detector tube or chemical indicator tube (see Chapter 4). This consists of a glass tube packed with a column of the chemical reagent impregnated on to a granular, inert, support material. Air is drawn through the tube and any of the specific contaminants in the air will produce a stain. The amount of the contaminant in the sampled air is estimated either from the length of the stain, as read off a calibrated scale printed on to the glass, or from measurements of the volume of sampled air required to produce a particular intensity of stain compared with a colour standard supplied with the tube.

Stain length tubes of two types are available:

(*a*) *Short-term tubes*[13] which give an instantaneous level reading, when air is drawn through the tube using a set number of strokes from a small hand-operated pump.

(*b*) *Long-term tubes*[14] which can be used for TWA monitoring where air is drawn through continuously at a low flow rate using a personal sampling pump over periods up to a full workshift. The stain length is calibrated in microlitres of vapour. This, when divided by the number of litres of air sampled, gives the TWA concentration of the contaminant in parts per million.

Specificity for a particular substance is obtained by using a reagent system which reacts selectively with that substance and no other. In practice it is difficult to achieve perfect specificity but interference can be minimised by incorporating a 'pre-cleanse' layer before the reagent, which selectively removes any interfering substances but not the substance to be determined. In general, specificity is highest for substances such as permanent gases which have unique chemical properties. For substances such as organic compounds, which are in the homologous series where the neighbours of a particular member of the series have similar chemical properties, the degree of specificity attainable is usually less than ideal. Detection limits for detector tubes are mostly above 1 p.p.m.

A further variant is to use a two-stage reagent system where, in the first stage, the substance reacts to form a different substance which then passes on to the second stage, forming the stain.

Examples:

Acetone – no precleanse layer, single reaction, concentration estimated by stain length, colour change: pale yellow to darker yellow.

Reaction: Acetone + 2,4-dinitrophenylhydrazine → yellow hydrazone.

Carbon monoxide – precleanse layer of Cr(VI) compound to remove hydrocarbons and H_2S, single reaction, concentration estimated by stain length. Colour change: white to brownish green.

Reaction: CO + iodine pentoxide (in the presence of selenium dioxide and pyrosulphuric acid) → iodine + carbon dioxide.

Benzene – precleanse layer of acid and aldehyde to retain interfering alkylbenzenes, two-stage reaction, concentration estimated by measurement of the number of sampling pump strokes required to give a stain intensity matching the comparison layer incorporated in the tube. Colour change: white to reddish brown.

Reaction: benzene + formaldehyde → diphenylmethane + sulphuric acid → *p*-quinoid.

1,1,1-Trichloroethane – two tube system, first tube reaction is oxidation to chlorine, second tube gives colour reaction. Concentration estimated by stain length. Colour change: pale grey to brownish red.

Reactions: oxidation tube converts the trichloroethane to chlorine; this passes to second tube containing *o*-dianisidine which reacts with the chlorine giving a brownish red reaction product.

Long-term hydrogen sulphide – simple single-stage reaction of H_2S with lead acetate. Colour change: white to brown. Sampling flow rate 10–20 cm^3/min, maximum sampling time 8 h. Concentration estimated by stain length calibrated in microlitres H_2S. Dividing by number of litres of air sampled gives H_2S in p.p.m. as a TWA over the sampling period.

3.2.2 *Paper tape monitors*[53]

These represent an extension of the detector tube principle where instead of air being drawn through the static reagent system it passes through a paper tape impregnated with the reagent, which moves across the sampling point at a constant rate.

The concentration of a particular substance is then estimated by the intensity of the stain developed. This technique dates back to the early 1950s when lead acetate papers were used for H_2S monitoring. In personal monitoring systems the complete

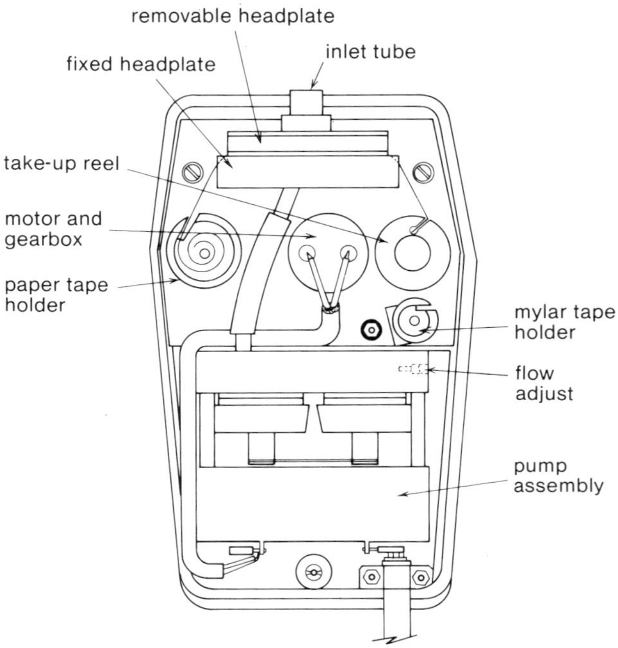

Fig. 5.1 Paper tape miniature continuous monitor[15] (MDA Series 4000). Shows MCM Unit with front cover removed.

tape is run through a personal sampler (Fig. 5.1) and the measurement of stain intensity is done afterwards.

Originally the intensity of the stain was estimated spectrophotometrically by measuring the light transmission through the paper. This limited the sensitivity and precision of the method due to variations in the paper and has since been abandoned in favour of measurements of the reflected light intensity.

In a similar way to detector tubes, a specific reagent system may be used which is impregnated on to paper, but in this case the reagent system is used dry. The paper is formed into a small reel or cassette and is incorporated into a midget personal sampler[15] shown schematically in Fig. 5.1, which can be worn by the subject for periods up to a full shift and powered by a separate battery pack.

In operation the paper is passed slowly between two plates to prevent longitudinal leakage of air. Air is drawn through an inlet hole in the centre of the plates and through the paper tape, using a constant flow pump and the exposed tape is then wound on to a take-up reel. Only one half of the tape is exposed, the unexposed half being used in the readout spectrometer as a baseline reference.

Fig. 5.2 shows a diagram of the readout spectrometer used for analysing the paper tape cassette after the end of the monitoring period. The tape is driven across an optical gate and light from a single source is split into two beams which are directed, via fibre optics, on to the exposed and unexposed portions of the tape at an angle of 90°. The light reflected at an angle of 45° is picked up by matched photocells and the

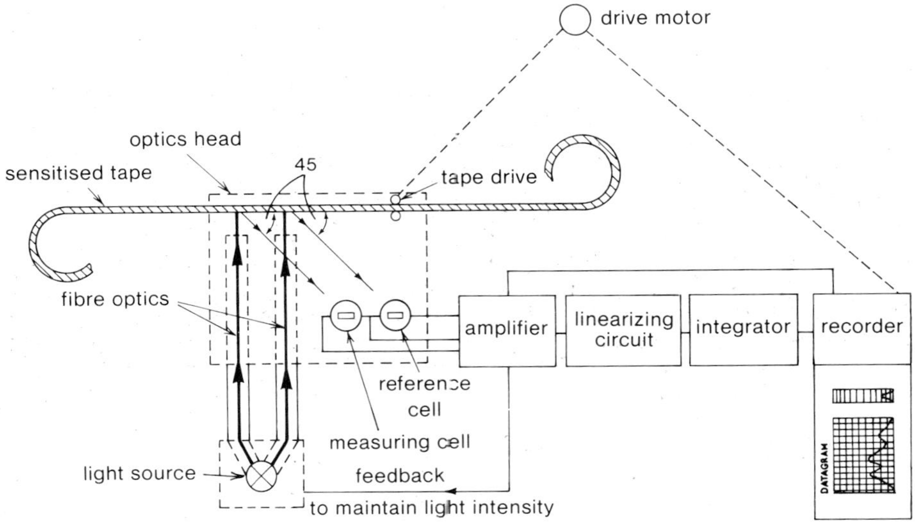

Fig. 5.2 Schematic diagram of paper tape readout[15] (MDA Series 4100).

difference between the two signals – which is proportional to the stain intensity – is recorded on a chart. This gives a recording of the exposure levels continuously throughout the monitoring period, the TWA concentration is also measured as an integration of the continuous level and marked on the chart at the end of the readout.

This system has considerable advantages over detector tubes as a continuous record of exposure is obtained and detection limits are much lower, i.e. within the range 0.001–0.5 p.p.m. The equipment costs are, however, considerably greater and the range of substances that can be monitored is more restricted, partly because the reagent systems must be used dry and must be stable in air, whereas detector tubes are kept sealed in glass, minimising deterioration and dehydration of the reagents. A portable version is also available, however, which incorporates the tape readout. Analysis of the tape is performed immediately after sampling giving a continuous though slightly delayed readout of concentration.

The substances for which tapes are available include: toluene diisocyanate, hydrogen sulphide, phosgene, nitrogen dioxide, chlorine, sulphur dioxide, vinyl chloride, hydrazine, toluene diamine, hydrogen chloride, ammonia, dinitrotoluene and organic isocyanates.

3.3 TWA sampling systems

The acceptability of breathing air quality where the toxicity of the contaminants is predominantly of a chronic nature is judged against TLVs usually set as 8-h time-weighted averages. Consequently the measurement techniques used should be capable of monitoring the air as close to the subject's face as possible and at a steady rate over the whole workshift.

The toxic effects of organic vapours which occur at the lowest concentrations are

Fig. 5.3 TWA sample collection by pump and gas bag. (Photo by courtesy of Sipin International, Inc.)

predominantly of a chronic nature and unfortunately these are the types of substance for which the direct reading equipment previously discussed is least suitable, usually on account of a lack of specificity, and in many cases, lack of sensitivity. In an industrial environment the substance to be measured is often present along with many other similar substances which could give interference on direct-reading sensors.

Because of these limitations the most commonly used monitoring technique for organic vapours consists of continuously sampling the subject's breathing air, usually collected at the lapel or collar. Samples can be collected either directly into a container or the contaminant substance can be removed by trapping on to an adsorbent or in a bubbler. The final sample, representing the average air composition over the monitoring period, is then analysed, usually in the laboratory.

Here, I will discuss the various types of sampling systems and their relative merits.

3.3.1 Slow filling containers

These collect the whole of the air at a constant rate without the use of adsorbents to remove the contaminant substances. The air sample is then analysed, usually by gas chromatography (GC). There are two main types used:

(*a*) *Gas bag and pump.* These comprise a low flow-rate personal sampling pump which draws air from the breathing zone and pumps it into a 1–2 dm^3 gas bag worn on the subject's belt, via a non return valve as shown in Fig. 5.3. The bag usually incorporates a septum to enable samples to be withdrawn for analysis using a gas syringe. Shorter period sampling can be achieved by using a higher pumping rate.

The gas bags used for this purpose must be impermeable to the contaminant substance to prevent losses due to permeation through the bag and adsorption by the bag lining. The most suitable types are made of inert plastics such as Mylar, Tedlar and PTFE and are often of laminar construction incorporating an aluminium layer to prevent permeation. It is desirable to test the rate of 'decay' in the concentration of the substance to be monitored to establish permissible sampling and storage times and any corrections for sample loss which may be necessary. Cross contamination between samples must be prevented by scrupulous cleaning after use to prevent any adsorbed vapour in the wall coating re-evaporating into subsequent samples. Cleaning is best carried out by purging the bags in a warm oven using a clean dry airstream for a few hours.

(*b*) *Evacuated metal containers.* These are small robust canisters, usually lined with an inert layer of PTFE to minimise adsorption, and have a critical orifice through which the air is sampled. In use they are evacuated to a measured pressure, the cap is removed for sampling and the container is filled at a rate controlled by the critical orifice which remains constant until a particular pressure is reached, after which it steadily decreases. This critical pressure defines the container's maximum sampling volume. Short-period samples can be taken using higher flow-rate orifices. After completion of sampling the pressure in the container must be measured in order to calculate the volume sampled; it is then filled to ambient pressure and the contents are analysed.

Both of these techniques are suitable only where the TWA level of the contaminant is well above the analyser's detection limit. This is more critical for the canister type where the sampled air is further diluted prior to analysis. The substances to be monitored must also be stable in air over the sampling period and must not be significantly adsorbed on to the container walls, or with the bag method, by the sample tubing and pump.

3.3.2 Impingers and bubblers

Gases and vapours can be absorbed from air into solvents or reagent solutions contained in a bubbler or impinger[16] through which the air is drawn at a regulated flow-rate using a sampling pump.

Bubblers are available which are small enough to be worn on the lapel and are designed to operate at various orientations. A typical bubbler and pump are shown at the top right of Fig. 5.4. The bubbles may be formed by the air being drawn through a narrow jet in the liquid (impinger) or through a glass frit to produce small bubbles. Some bubblers may contain baffles or glass beads to slow the passage of the bubbles through the liquid and increase the gas/liquid contact time (see Chapter 4).

For many contaminants, the extraction efficiency of the trapping solvent is seldom 100% as the solute will often have a finite vapour pressure above the solvent, which will increase with its concentration in the liquid. Solvent evaporation will also occur from volatile liquids.

It is preferable to use an extraction liquid containing reagents which react with the contaminant to form a non-volatile product rather than relying on simple solubility of the contaminant in the liquid (see Chapter 4).

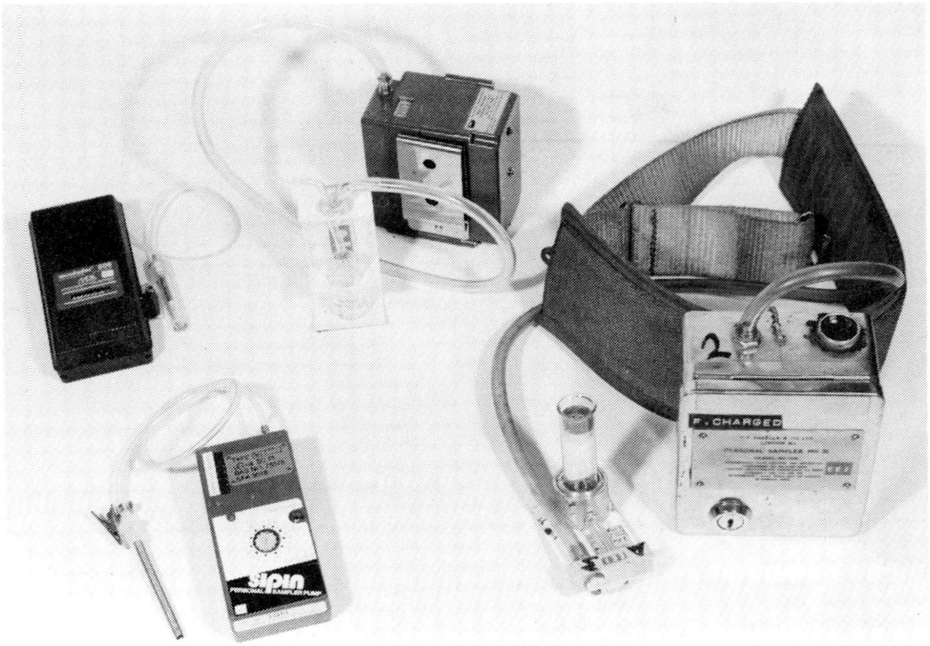

Fig. 5.4 Personal sampling pumps and vapour adsorption devices. *Right:* older type silica gel adsorption tube; *top:* midget impinger; *left top:* NIOSH type tube with low-flow personal sampling pump; *left lower:* heat desorbable vapour adsorption tube with low-flow personal sampling pump.

Examples:

Hydrazine – absorbed in phosphomolybdic acid solution and analysed spectrophotometrically.[17]

Lead alkyls – absorbed in iodine/methanol solution and analysed colorimetrically.[18]

In general impingers/bubblers are inconvenient to use, rather clumsy to wear, require large, high-flow rate pumps and do not allow very low detection limits to be achieved. They are most useful for sampling substances which are unstable or reactive and cannot be recovered from solid adsorbents either because they are very strongly adsorbed or because they react, polymerise or are oxidised on the surface.

Another application where their use may be preferred is for vapours from high boiling point substances which may be difficult to remove from solid adsorbents.

3.3.3 Vapour adsorption tubes

The most widely used, and probably the most convenient technique for collecting TWA samples of gases and vapours is the vapour adsorption tube. This is a metal or glass tube containing a bed of adsorbent through which air is drawn at a regulated flow rate, using a personal sampling pump. Any contaminant vapours in the air are

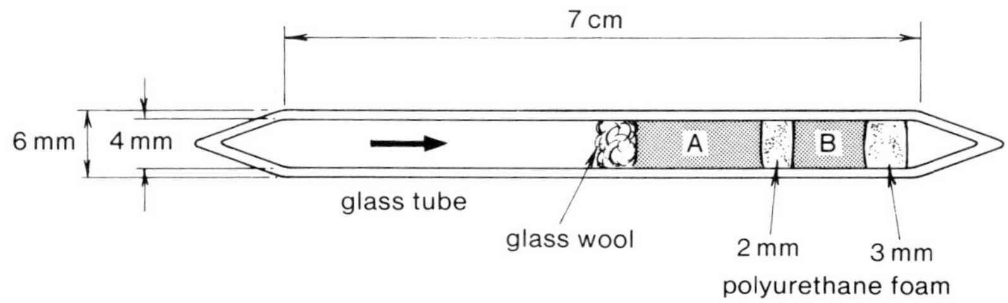

Fig. 5.5 NIOSH type vapour adsorption tube. A, 100 mg; B, 50 mg of 20/40 mesh charcoal (coconut husk).

trapped and held on the adsorbent. They are then analysed by first desorbing the trapped vapour, either using solvent extraction, or heating and flushing with an inert gas, and then injecting the solvent or flushing gas into an analyser, usually a gas chromatograph.

This desorption step is a crucial stage in the technique as the trapped vapour must be quantitatively removed without degradation and with a minimum of dilution to maximise sensitivity, and in a form compatible with the analyser.

Many types of adsorbent and traps have been used, including silica gel, alumina, molecular sieve and charcoal. An early form of sampling tube using silica gel is shown in the bottom right hand corner of Fig. 5.4. These tubes often used several grams of adsorbent and required solvent desorption to remove the sample. This restricted the choice of adsorbent to inert inorganic materials and required large volumes of air to be sampled to achieve adequate sensitivity. Large, heavy sampling pumps were therefore necessary which were cumbersome to the wearer as were the large sample tubes and holders. A further problem was that, with the exception of charcoal, the adsorbents previously mentioned would react chemically with water. This made careful preparation necessary to achieve a consistent initial adsorbent activity. It also made the prediction of the volume of air that could be sampled without the substance breaking through the bed (break-through volume) very variable and dependent on the moisture content of the air.

These early forms of tubes have now been largely replaced by miniature samplers and low flow rate pumps illustrated on the left of Fig. 5.4. They still find some use for sampling highly polar substances that may not be readily desorbed from charcoal.

The NIOSH[19] tube

The first standardised miniature adsorbent tube to find widespread use as the basis of specification monitoring methods[21] was the NIOSH tube. This consists of a glass tube containing two beds of charcoal as shown in Fig. 5.5.[20] The tubes are sealed until required for use, when the tips are broken off for sampling. Air is drawn through the tube on to the larger bed first, the second bed being used to detect any breakthrough from the first.

The advantages that this tube gave were in the use of a standardised grade of coconut husk charcoal, which is a very strong adsorbent for organic vapours. This allows the sample to be collected on a very small bed and to be desorbed with a small volume of solvent, typically 0.5–1 cm^3, thus minimising dilution and giving good sensitivity. Sampling volumes are also low, allowing the use of much smaller personal sampling pumps which can be worn inside an overall pocket and are much less obtrusive to the wearer, as is the small sampling tube which is usually inserted into a holder for clipping on to the collar (see top left of Fig. 5.4).

The limitations on sampling flow rates and periods have now been defined for many substances along with standard sampling and analysis procedures.[21,23–25]

After sampling, the tubes are stopped with plastic end-caps prior to analysis. Analysis involves cutting open the tube and carefully removing both adsorbent beds into separate glass vials. These vials have 0.5 cm^3 or more of a solvent added (usually carbon disulphide, methylene chloride, etc.), and they are shaken and left for a set period. The solvent is then made up to volume, mixed and aliquots are injected into the analyser.

Although this tube was a considerable improvement over previous types, there are still many drawbacks connected with its use:

(*a*) Sample preparation is time-consuming and uses toxic flammable solvents which require precautions to be taken. Laboratory facilities are necessary for this analysis.

(*b*) Sensitivity is still limited by the solvent dilution step. Use of a minimum of 0.5 cm^3 of solvent and a maximum sample injection of 5 μl gives a minimum sample dilution of 100 times.

(*c*) Solvents often contain impurities which may interfere in the analysis. The presence of a large bulk of solvent also complicates the analysis.

(*d*) Recovery is often incomplete as the equilibrium between the species to be desorbed and the solvent to replace it may not lie completely on the solvent side. Table 5.1 shows recoveries quoted for ninety-six substances.[22]

(*e*) Sample tubes cannot be reused.

(*f*) The sample is recovered in the liquid phase, whereas most analysers work in the vapour phase.

With the ever-increasing trend toward lower TLVs and the increase in monitoring for environmental levels of contaminants outside the workplace, the sensitivity limitations of the NIOSH-type tube have led to an increase in sampling volumes and

Table 5.1 Sample recovery from charcoal sampling tubes

Percentage recovery	*Number of substances*
100	9
95–100	30
90–95	24
85–90	16
80–85	7
75–80	8
70–75	2

flow-rates being used. Special clean-up procedures have to be used for removing any traces of contaminants on the charcoal prior to packing the tubes and for further purifying the solvents used for desorption.

The charcoal used in these tubes is a fairly coarse mesh size (20/40) and the bed diameter is kept large to minimise flow restriction at higher flow rates. These factors, coupled with the small size of the bed, limit the breakthrough volumes obtainable, further restricting sensitivity. Simple tests have shown[26] that, if the adsorbent particle size is decreased to 60/85 mesh BS, the breakthrough volume is increased by more than twice. Furthermore, with the same amount of adsorbent, if the bed diameter is reduced from 4 to 2.5 mm, the breakthrough volume is increased by more than three times. These effects are predictable as the breakthrough process is similar to frontal elution gas chromatography and the effects of smaller particle size, increased bed length and decreased bed width all result in an increase in plate efficiency of the column.

Heat desorption[26–27]

An alternative to solvent desorption samplers, which is now gaining rapidly in popularity, is the heat desorption technique where the adsorbent is flash heated and flushed with an inert gas. The desorbed vapour is carried away in the flushing gas and either introduced directly into the analyser or stored in a reservoir. This technique overcomes most of the disadvantages of solvent desorption.

(*a*) Sample preparation is eliminated.

(*b*) No solvent or impurities from the solvent to interfere with the analysis.

(*c*) Sensitivity is much greater as the whole sample can be analysed so that lower sampling flow rates and longer sampling times can be used.

(*d*) Recovery is generally more complete.

(*e*) Sample tubes are clean and ready for re-use after each analysis.

(*f*) The sample is recovered in the vapour phase allowing a wider range of analysers and detectors to be used.

(*g*) Portable analysers are available eliminating the need for a laboratory and allowing on-the-spot results to be obtained.

The bed size can be larger with this type of tube and typically, five times more adsorbent can be used. This allows longer sampling times to be achieved without the risk of breakthrough. Organic adsorbents, such as resins, porous polymers and GC column packing materials, can be used, increasing the flexibility of the technique.

A typical heat-desorbable vapour adsorption (VA) tube is shown in Fig. 5.6.

These tubes are constructed from stainless steel tubing, usually with thin walls to aid heat flow, marked at one end to indicate that used to collect the sample and numbered for identification. Types of adsorbents used include charcoal for permanent gases, and porous polymer for vapours.

The function of the thermal desorber is to quantitatively remove the trapped

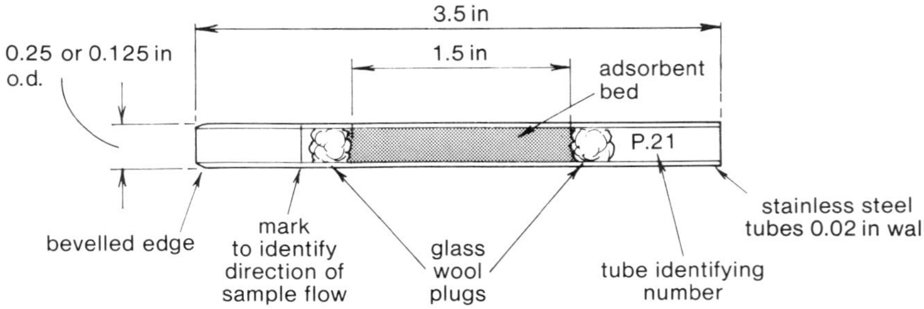

Fig. 5.6 Heat desorbable vapour adsorption tube.

vapour, without degradation, and transfer it to the analyser with the minimum of dilution in the flushing gas.

There are several types of heat desorber, the simplest being attached to a gas chromatograph and tapped into the carrier gas line. To operate it, the VA tube is mounted and flash-heated and the carrier gas is diverted through the tube. The desorbed vapour is then carried into the GC as a concentrated slug in the carrier gas and analysed directly, using an isothermal separation column and usually a flame ionisation detector. A variation of this type of desorber incorporates a cold trap after the desorber to retrap the vapour, which because of its low thermal capacity can be very rapidly flashheated so that the vapour enters the analyser as a very short pulse. These are both called direct desorbers, they have the advantage of simplicity, and are quick and easy to use. Analysis time is minimised by using the GC analyser in the isothermal mode. With a well-designed desorber and the correct type of VA tube, the direct desorption technique without the second cold trap is suitable for the greater majority of applications as the small amount of bandspreading during desorption results in only minor peak broadening in the analyser. In situations where peak broadening interferes with the separation, the problem can often be overcome by desorbing on to a cold column and then temperature-programming up to the operating temperature.

There are also indirect types of desorber where the desorbed vapour is collected in a reservoir or large gas syringe. Aliquots from the syringe can be analysed by flushing into the loop of a gas sampling valve and then injecting into the analyser. The indirect system makes it possible to make replicate analyses or analyses on different analysers. It is also useful when a high resolution GC separation is necessary, as with capillary columns. Some sensitivity is lost using the indirect technique and the equipment is more complex and expensive.

Because the desorbed vapour is released in the gas phase and contained in a small volume of inert gas, heat desorption is ideal for use with a wide range of analysers. In some cases analysis can be carried out directly using the detector tubes in the carrier gas outlet line. Some other examples are:

Atomic absorption spectrometers for stable volatile organometallic compounds such as lead alkyls.[38] The desorber gas is fed directly into the burner just under the slot with the spectrometer set up using a recorder on the output as in FAAS.

Long path gas-cell infra-red spectrometers can be used where GC separation is difficult, particularly for analysis of a compound with a functional group different from those in other compounds present which gives a discrete absorption band, separated from the absorption bands of the rest of the compounds, e.g. organic halogen compounds in the presence of hydrocarbons.

Total hydrocarbon meters: flame ionisation detectors, photoionisation detectors, reaction coulometers, explosimeters, etc. These detectors may be used where it is known that only one substance is present or where the concentration of total gaseous organic compounds is required.

Limitations of vapour adsorption tubes[26,39,54]

The performance characteristics required for such tubes cover many parameters, including:

(*a*) Sufficiently high breakthrough volume to allow full shift sampling periods to be covered.

(*b*) Sufficiently high loading capacity to cope with the highest levels likely in the sampled air.

(*c*) Breakthrough volumes not affected by contaminant concentration or by atmospheric humidity.

(*d*) Retention of sample during storage without loss or degradation.

(*e*) Complete desorption of sample without any degradation and without any 'memory' from previous samples.

Breakthrough volume

This is defined as the volume of air that can be sampled before the substance breaks through the adsorbent bed. It is one of the most important factors since, for the determination to be valid, no loss from breakthrough must occur. The NIOSH tube relies on the use of a back-up adsorbent bed to detect any breakthrough from the first bed. This is not a very satisfactory approach as it doubles up on the analysis time and can lead to wasted samples. The best approach is to be able to predict confidently the breakthrough volume for any substance at the particular sampling temperature so that sampling times and flow-rates can be adjusted to prevent breakthrough occurring.

The breakthrough volume is dependent on many factors, mainly the surface area and activity of the adsorbent, its mesh size, the dimensions of the bed and the temperature. It is independent of flow-rate between certain limits (typically 5–80 $cm^3 min^{-1}$ through a 0.25 in diameter VA tube) and also independent of humidity up to near saturation point.

Vapour concentration affects breakthrough volume for some adsorbents more than others. Charcoal is almost unaffected up to its saturation capacity, whereas polymers show a small concentration dependence. This is illustrated in Fig. 5.7, which shows plots of breakthrough volume against total vapour adsorbed up to the breakthrough point for two adsorbents. Curve A is for benzene vapour at 50 °C on a

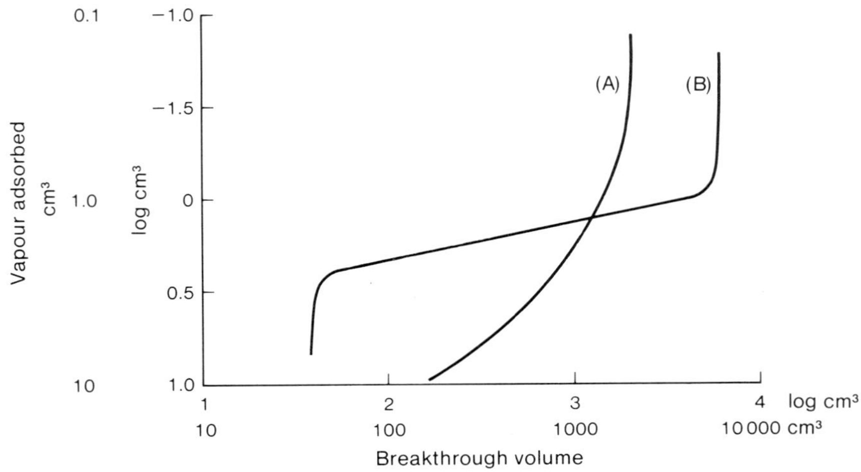

Fig. 5.7 Breakthrough volume versus volume of vapour adsorbed. A, Benzene in air at 50 °C on 0.25 in VA tube – 300 mg Poropak 'Q'; B, acrylonitrile in air at 50 °C on 0.125 in VA tube – 70 mg charcoal.

6 mm o.d. VA tube containing a 300 mg bed of porous polymer adsorbent (Porapak 'Q', 80/100 mesh) and curve B is for acrylonitrile vapour at 50 °C on a 3 mm o.d. tube containing 70 mg NIOSH grade charcoal (50/60 mesh). These curves reflect the shape of the adsorption isotherms, which, if known, can be used to predict the degree of concentration dependence on breakthrough volume for an adsorbent as well as the loading capacity.

For a standardised VA tube with fixed dimensions, adsorbent type, mesh size and weight, used within its loading capacity, sampling air which is below saturated humidity and within its flow rate range, the only factors left which will affect breakthrough volume are: (*a*) the nature of the substance, basically its heat of adsorption on the particular adsorbent which is a function mainly of its volatility and chemical type; (*b*) temperature, with the log of the breakthrough volume being directly proportional to the reciprocal of absolute temperature in the same manner as retention volume depends on temperature in gas chromatography.

These parameters can be reduced to a single equation for a standardised VA tube, relating the breakthrough volume to temperature and the 'effective' boiling point of the substance, an example being:

$$\mathrm{Log}\, R_v = (E \times 12.69 - 763.3)(1/T) - (E \times 0.00908) + 1.278$$

for a 0.25 in o.d. VA tube containing 550 mg of a NIOSH grade charcoal where: R_v is the breakthrough volume in cubic centimetres, T is the temperature in degrees Kelvin and E is the 'effective' boiling point in degrees Kelvin.

This equation was derived experimentally by measuring breakthrough volume at various temperatures to obtain a series of plots as shown in Fig. 5.8. The breakthrough volume plots above 3–4 dm^3 are extrapolated.

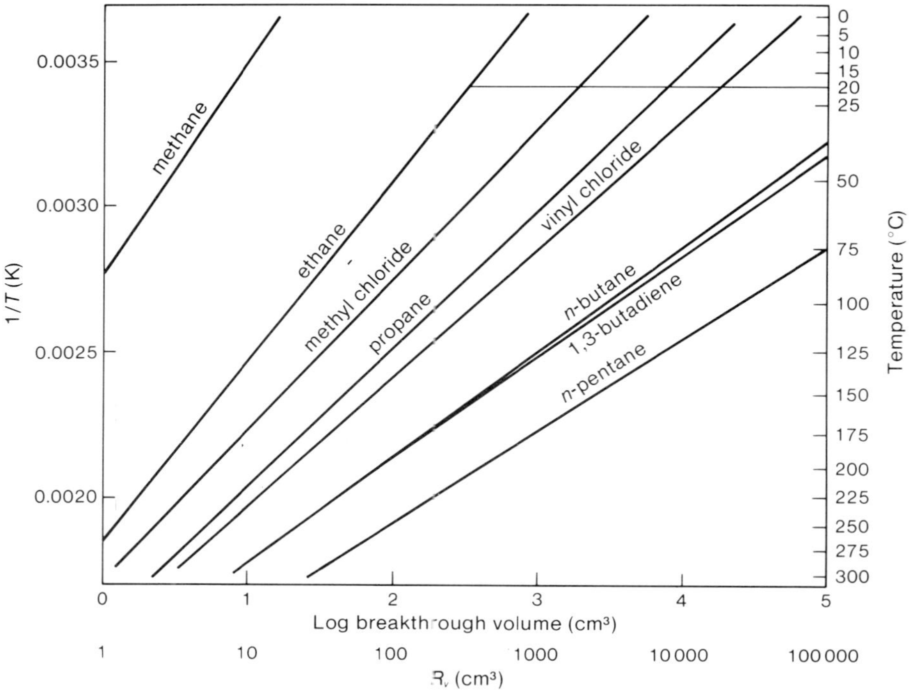

Fig. 5.8 Breakthrough volumes for 0.25 in charcoal tubes (550 mg) versus temperature.

Effective boiling point (EBP) is used as little data are available on heats of adsorption and this quantity is in any case difficult to measure. EBP is a more convenient parameter to use and is easy to measure; it is the boiling point of the substance indicated by its chromatographic retention time through a column of the absorbent, relative to *n*-alkanes which are used as reference.

There is quite a lot of published information on retention times – either absolute times or as Kovats or other retention indices from which the EBP can be estimated. Retention times can also be easily measured by setting up a VA tube as a GC column at an elevated temperature so that peaks from small injections of vapour are eluted in 5–10 min. The retention times for the unknown substance, and for *n*-alkanes selected to bracket it, are then measured and the EBP of the substance is read off from a plot of boiling point against retention time of *n*-alkanes.

Desorption

With solvent desorption, recovery is often less than 100% as shown in Table 5.1, with the most polar or reactive substances being usually the least efficiently desorbed. This factor can be controlled if the desorption procedure is carefully standardised by using the same extraction time, temperature and degree of agitation, and by using VA tubes from the same batch. This standardisation helps to give a

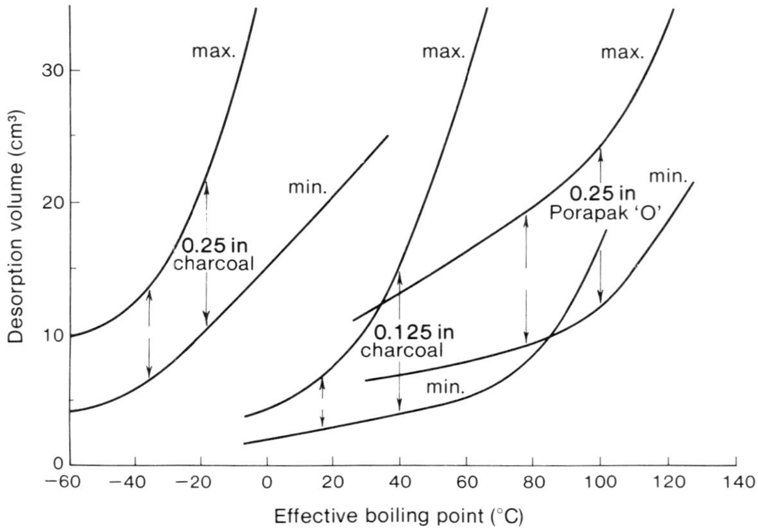

Fig. 5.9 Desorption volumes.

repeatable desorption efficiency, which can be measured and allowed for in the calibration procedure.

The efficiency of heat desorption is usually close to 100%, losses occurring only by polymerisation, oxidation, thermal cracking or irreversible reaction of the adsorbed substance with the adsorbent. These processes occur only with very reactive or unstable substances and for the great majority desorption is complete. Obtaining a satisfactory analysis from direct heat desorption into an isothermal GC requires that the substance be desorbed into the smallest possible volume of flushing gas in order to prevent excessive peak broadening.

The technique works best using 0.25 in o.d. packed columns with peak retention times of not less than 3 min in which case desorption volumes of 30–40 cm^3 will only broaden the peaks about one-and-a-half times. Fig. 5.9 shows a plot of desorption volume against effective boiling point for two types of VA tube. The desorption volume is not a fixed quantity but varies from a minimum value, when a small amount of vapour is adsorbed on to the end of the adsorbent bed, to a maximum when a larger amount of vapour is adsorbed along the full length of the bed. The increase in desorption volume with EBP limits the range of substances for which a particular type of tube can be used, and so more than one type of adsorbent is necessary to cover the range from gases of low b.p. up to liquids of high b.p. The useful range for a tube is higher than is indicated in Fig. 5.9. The reason is that a substance with a high desorption volume will also have a high breakthrough volume and so, in practice, does not penetrate far along the bed, giving desorption volumes nearer to the minimum rather than the maximum values. The selection of the most suitable type of VA tube for a particular substance is then based on the best compromise between a high enough breakthrough volume to enable full shift monitoring and a low desorption volume to give minimal peak broadening. In practice this range can be covered with two or three types of VA tube, typically with charcoal or porous polymer adsorbents.

3.3.4 Passive samplers[40–46]

Principles

Solid absorbents can be used to sample contaminant vapours in air without the use of pumps to draw the air through the adsorbent. A layer of adsorbent, when exposed to air, will collect contaminant vapour at a rate determined by the rate of diffusion of the vapour through air. An adsorbent layer is usually incorporated in a container with a layer of stagnant air of fixed depth maintained motionless by a draught shield (Fig. 5.10). Under these conditions the vapour in the air above the adsorbent will be removed from the air on to the surface of the adsorbent resulting in a concentration gradient across the stagnant air, which varies from essentially zero at the adsorbent to the original concentration in the outside air. This concentration gradient will produce a flow of the vapour across the stagnant layer due to diffusion, and the rate of flow will be a function of: (i) the diffusion coefficient, $D(\text{cm}^2\,\text{s}^{-1})$; (ii) the concentration gradient $C-Ca$ (mol cm^{-3}) (where C = vapour concentration in the outside air and Ca = vapour concentration above the adsorbent); (iii) the diffusion gap, $L(\text{cm})$; (iv) the cross sectional area of the collector surface, $A(\text{cm}^2)$.

According to Fick's Law of diffusion,

$$\text{Uptake rate}\,(\text{mol s}^{-1}) = \frac{D \cdot A(C-Ca)}{L}$$

For a sampler of fixed dimensions the uptake rate for a particular substance is directly proportional to ambient concentration and so the amount collected over a given time will be proportional to the average concentration over this period.

Various types of passive sampler have been developed.[40–45] Some rely purely on a diffusion gap for controlling the uptake rate, and others have a plastic membrane in place of the diffusion gap to control uptake rate, which is then dependent on the rate of permeation of a substance across the membrane; this is also directly proportional to concentration gradient. However, permeation rates through plastics are markedly temperature-dependent and to reduce this, porous plastic membranes have been introduced in which the diffusion is essentially through the gas phase in the capillaries of the membrane. Other designs make use of a diffusion gap plus a membrane.

Most of these open-faced types of passive samples have a fairly large collection area ($5\ \text{cm}^2$) as the analysis has to be made by solvent-extracting the collected vapour from the adsorbent layer for subsequent GC analysis. This large collector area is necessary as uptake rates are relatively low and much sensitivity is lost owing to dilution of the sample in the extraction solvent. The effective sampling rate is about $50\ \text{cm}^3\,\text{min}^{-1}$ and the sampler is thus only applicable for fairly long period sampling. This type of passive sampler has a number of inherent disadvantages, such as:

(*a*) Sealing of the sampler, when not in use, to prevent contamination.

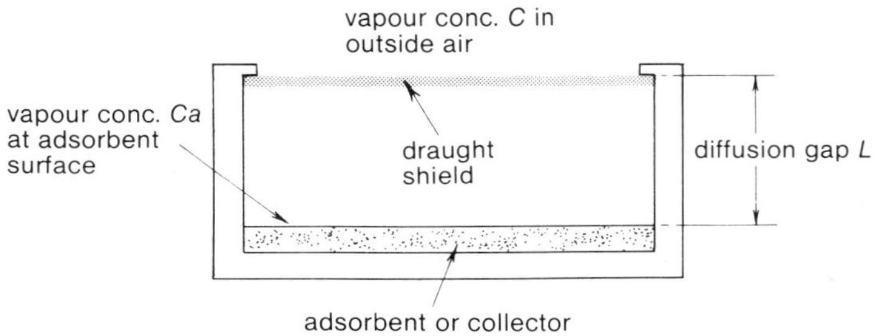

Fig. 5.10 Diagram showing the principle of a passive sampler.

(*b*) Preventing contamination of the adsorbent layer after sampling when it is removed for analysis and exposed directly to the laboratory air.

(*c*) Rather high preparation and analysis time in analysing and recharging the adsorbent layers.

Most of these problems have now been largely overcome and one type is designed so that solvent extraction can be carried out inside the sampler itself without the adsorbent being removed. This sampler is also disposable.

A recent development is the tube-type heat desorbable passive sampler[46,47] which is basically a 0.25 in o.d. heat desorbable VA tube in which the adsorbent bed is accurately positioned at a fixed distance from one end of the tube (typically 1–2 cm), which is closed by a thin membrane and carried in a holder. The other end of the tube is sealed off. Vapour permeates across the membrane and along the tube to be collected on the adsorbent bed giving an effective sampling rate of about 0.5 $cm^3\ min^{-1}$. Although the collection area is relatively small, this is more than compensated for by the increase in sensitivity gained from direct heat desorption which eliminates the sample dilution inherent in solvent extraction. The possibility of contamination during handling and storage is minimal with these types of sampler and sample preparation and refurbishing time are almost eliminated. Moreover, the sampler can be re-used many times without changing the adsorbent.

Many manufacturers now market passive samplers with the cost of the analysis included, so that the customer exposes the sampler, seals it and posts it back to the supplier, who analyses it and returns the refurbished sampler to the customer along with the results. This type of service is likely to be most attractive to small industries who cannot afford their own facilities.

Much of the evaluation work on passive samplers to date has been for single substances or two-component or three-component mixtures in the laboratory at steady concentrations, and under these conditions their performance has been satisfactory. However, field data are now becoming available on comparisons with pumped tubes which shows that there is often a systematic bias. Agreement between pumped and passive samplers is seldom better than a relative standard deviation of 20%.[68,69]

Advantages and disadvantages

The advantage of passive samplers over the pump and tube samplers is solely in the elimination of the pump. This makes the sampler lighter and smaller and so more acceptable to the wearer. In addition the cost is reduced and pump reliability problems are overcome.

There are some disadvantages:

(*a*) The uptake rate of each substance must be measured and calibrated in a laboratory. The results must be assumed to apply to field conditions until side-by-side field trials in which pumped and passive monitors are compared directly can establish the correlation.

(*b*) The fixed uptake rate allows no flexibility in sampling rate to accommodate short-term operations.

(*c*) The low uptake rate restricts the suitability of these samplers for use at very low concentrations and short periods.

There is no breakthrough limitation on passive samplers but it is possible to saturate the collector. Thus a knowledge of its loading limitations is necessary, so that the maximum amount of vapour collectable is not exceeded.

Sensitivity to ambient conditions

Several ambient factors can affect diffusion and permeation rates, including temperature, pressure, humidity and air velocity and some correction to uptake rates may need to be applied.

(*a*) Temperature and pressure variations affect the diffusion coefficient in the manner:

$$D = \frac{T \cdot K^{3/2}}{P}$$

giving a theoretical temperature coefficient of around 0.6% $°C^{-1}$ and a pressure coefficient of around 1.3% $(\text{cm Hg})^{-1}$.

(*b*) Humidity effects are not predictable but variations have been found experimentally with certain types of sampler, particularly at extremes of humidity.

(*c*) Air velocity effects over the face of the sampler have been found to affect some types of passive sampler, particularly those with deep wells and obstructions on the fact which can cause areas of stagnant air and decreased uptake rates.

Fundamental sources of error

One fundamental source of error arises from the time constant characteristic of the passive sampler when used in atmospheres with rapidly fluctuating concentrations. In many industrial exposure situations where a person works around a localised source of release the concentration varies rapidly with time. Experience[3] has shown that exposure is then due mostly to peaks, often of high concentration but short duration.

Most passive samplers have relatively large time constants, i.e. the time taken for a contaminant molecule arriving at the outer face of the sampler to traverse the diffusion gap or membrane and be adsorbed by the collector.

$$\text{The time constant (s)} = \frac{L^2\,(\text{cm}^2)}{D\,(\text{cm}^2\,\text{s}^{-1})}.$$

So for a sampler with a 1 cm diffusion path the time constant for benzene vapour will be around 11 s. A pulse of vapour may enter the sampler, but if the outside concentration falls, a concentration gradient will exist in two directions, on to the adsorbent and also back out of the sampler, and so some vapour will be lost by diffusion back out of the sampler.

Evaluation work on passive samplers has so far been largely at steady concentrations. Theoretical considerations indicate that pulses of vapour with durations less than about twice the time constant of the sampler will not be quantitatively collected.[68,70,71]

Calculations and experiments have shown that with substances that are not strongly adsorbed there will be a significant equilibrium vapour pressure of the substance above the adsorbent which will decrease the sampler uptake rate as can be seen from the equation at the beginning of this section, describing Fick's Law of Diffusion. The magnitude of this effect is only significant on charcoal for very light gases, but with the porous polymer bead adsorbents which are used in the heat desorbable passive samplers (because vapours from liquids are too strongly adsorbed on charcoal to be satisfactorily heat-desorbed) significant uptake rate losses have been noted for vapours from liquids with boiling points up to 80–90 °C.

Theoretical predictions show that the decrease in uptake rate will not be constant but will vary with concentration and will also be affected by fluctuations in concentration. This is because the rate at which the adsorbed species is removed from the face of the adsorbent bed by migration into the bed is very slow compared with the time constant of the sampler.[68]

3.3.5 Body fluid monitoring

When vapours are inhaled, a proportion passes across the lung membrane into the bloodstream and is transported to the organs and body tissue. Most water-insoluble organic substances accumulate in the fatty tissue, liver, kidneys, etc. The reverse process also occurs where the substance is eliminated in the expired air, and under steady exposure conditions an equilibrium will eventually be reached at which the uptake and elimination rates are the same. A proportion of the substance in the body is usually also eliminated via metabolic reactions in the organs and excreted as one or more metabolites in the breath, urine, faeces or sweat. When exposure ceases, the substance continues to be eliminated at a rate which decreases exponentially with time.

Analyses of body fluids are used as a measure of the body burden of the substance and, although these do not necessarily correlate precisely with time-weighted average personal air monitoring levels, the body fluid concentration has, in a few cases, become an acceptable measure of exposure, with maximum allowable levels established in a similar manner to TLVs.

Common applications are:

(*a*) *Urine analysis*, for metabolites such as trichloroacetic acid from trichloroethylene exposure[48] and phenols from benzene exposure.[49,50] The samples must be taken at a particular time after the end of the exposure period and the measurement must be compensated for variations in the body's metabolic and water elimination rates by correcting the result either to a standard SG or a standard creatinine content.

(*b*) *Blood analysis*, usually for substances which are eliminated from the body very slowly, such as lead,[51,52] cadmium, chromium, etc., samples being commonly taken from a small prick in the ear lobe. This method is used only occasionally, as blood sampling is unlikely to be acceptable on a routine daily basis.

(*c*) *Analysis of exhaled breath*[55–62] taken after exposure ceases, for levels of vapours and gases eliminated via the respiratory system. This is a technique which is attracting much interest as a means of replacing both urine analysis and personal monitoring. A common example is the dreaded breathalyser test for alcohol in motorists.

An extension of these techniques, which is less than satisfactory for obvious reasons, is medical monitoring for the detection of symptoms of occupational diseases such as lung function tests for chemically induced asthma and chest X-rays for pneumoconiosis from exposure to silica and asbestos dusts etc. The use of the person as a monitoring system, where the assessment of exposure is based on measuring the degree of resultant damage, is wholly unacceptable as a personal exposure monitoring technique and cannot be condoned.

3.3.6 Future developments

The technology of air quality monitoring is a fast developing field and yesterday's new technique can become tomorrow's obsolete one in a surprisingly short time. Current techniques are always being developed, extended and improved and new ideas and instrumentation are common.

In the near future, the range of applications for paper tape monitors and detector tubes is likely to be extended but it is unlikely that these devices can be made specific enough for a very comprehensive range of individual organic substances without interference from their homologues.

Passive samplers are likely to become a more common and accepted technique with an increase in the number of manufacturers and suppliers offering an analytical service for both passive and pumped samplers. The tube-type heat desorbable sampler will probably be more popular for those with their own analytical facilities and automated desorption equipment will soon be available.

Exhaled breath analysis is another technique which seems likely to attract increasing interest as the analytical technology is now available to enable rapid, specific and precise measurement to be made of traces of contaminant vapours eliminated in the breath subsequent to exposure (Fig. 5.11).

The attractions of the technique are the elimination of prolonged personal sampling, as a breath sample can be collected in less than 5 min and analysed quickly and easily, giving a measure of the person's average body burden of the contaminant arising from exposure over several previous workshifts.

Fig. 5.11 Sampling exhaled breath. Inhaled air is cleaned through a charcoal filter, then exhaled air diverted into gas bag collector. (Photo by permission C. Pollard.)

Contaminants in the body are eliminated both in the exhaled breath and by metabolism in the body organs. Initially the elimination rate is rapid but it decreases exponentially with time and after several hours the change in elimination rate is slow and the concentration in the exhaled breath reflects the longer term body burden.

In practice samples of breath can conveniently be taken only when the person starts the next workshift – usually 16 h after the end of the previous one. Samples taken at the end of an exposure period would be representative only of very recent exposure levels.

Sampling technique is important as the maximum concentrations will be found in the air in immediate contact with the lung membrane, the alveolar air. Air in the higher regions of the lungs and the bronchial tubes will contain much lower concentrations. This factor has resulted in two different approaches. One is to try to isolate and collect the alveolar air by sampling only the last part of the breath from a single exhalation. The other approach is to sample the whole breath averaged over several breaths to obtain a measure of the actual elimination rate. Sufficient evidence is not yet available to indicate which approach will give the best results.

The main problem with exhaled breath analysis is that present standards, i.e. TLVs, are for concentrations in the inhaled air, and no standards exist as yet for exhaled air. To validate the technique, the usual approach is to try to correlate exposure with exhaled air levels, whereas logically the direct measurement of the body burden itself should be the preferred standard.

The magnitude of a toxic effect depends on the concentration of the damaging substance in the critical organ, which will be related to the general body burden. This is, however, dependent not only on the concentration in the inhaled air, but also on several other variable factors:

(1) Respiration volumes can vary two to three times between sedentary work and vigorous work.

(2) The degree of bodily absorption of a substance from the inhaled air will differ according to exposure patterns. A steady concentration can reach an equilibrium where absorption and elimination rates are the same. Much actual exposure in practice, however, occurs as a series of peaks and troughs, and so no equilibrium is likely to occur and furthermore the percentage of the substance absorbed from the air may vary with its concentration.

(3) The rate at which substances are metabolised and eliminated from the body can differ from person to person, can be affected by the presence of other contaminant substances[72] and may depend on the amount of fatty tissue available for storage of water insoluble substances.

(4) Absorption can occur by routes other than inhalation such as orally or dermally.

It is therefore evident that standards for acceptable levels in exhaled breath should not be related to TLVs. Until such a different philosophy becomes acceptable, it is unlikely that this technique will be used to its full potential, which is both to simplify monitoring and to improve the measurement of the body burden of the substance.

The pocket dosemeter/continuous meter is probably the most attractive instrumental development. Ideally this comprises a selective detector with solid state electronics incorporated into a case small enough to be worn in the top pocket of an overall. The signal from the detector could be displayed on a small meter or digital display as concentration, with a hazardous level alarm to sound a warning when a preset level is exceeded. An integrating circuit could store the cumulative dose electronically and this would be measured after the end of a shift via a readout device. A further facility could be the provision for connection to a miniature recorder to store a continuous recording of the contaminant level.

The electronics technology for such a monitor already exists, but the problem area is the detector, which would need to be selective, sensitive and stable in both its zero setting and its calibrated response. It remains to be seen if electrochemical, solid-state or other sensors can be perfected sufficiently to make personal monitors of this type suitable for a significantly wider range of substances than those presently available, which are mostly for inorganic gases.

4 References

1. *Occupational Exposure Sampling Strategy Manual.* US Dept. of Health Education and Welfare. NIOSH Publication No. 77–173 (1977).
2. R. J. Sherwood. *Amer. Ind. Hyg. Assoc. J.* **27**, 98–109 (1966).
3. S. J. Silk. *Legal and legislative aspects of toxic hazards.* SIRA Conference on Health and Safety Monitoring of Toxic Gases, Vapours and Dusts. London (1979).
4. European Standard EN50 020. *Electrical Apparatus for Potentially Explosive Atmospheres.* Part 1, *General Requirements,* part 7, *Intrinsic Safety* (1977).

5. Originally British Approvals Service for Electrical Equipment in Flammable Atmosphere (BASEEFA) Certification Standards SFA 3012: 1972. New standard is EEC standard issued under British Standard Institution BS 5501 (1977).
6. M. A. Daynes. *Gas Analysis by Measurement of Thermal Conductivity.* Macmillan, London (1934).
7. A. R. Baker and J. G. Firth. The estimation of firedamp: applications and limitations of the pellistor. *Min. Eng. No. 100,* pp. 237–44 (1969).
8. A. W. Brewer. *New Scientist,* July 1957, p. 32.
9. P. Hersh. British Patents 707 323 (1953), 750 254 (1954).
10 L. C. Clark. *Trans. Amer. Soc. Artif. Lit. Organs.* **2**, 41 (1956).
11. N. Taguchi. British Patent 1 280 809 (1970).
12. Dries, R. L. Silicone semiconductor gas monitors. *Int. Environ. Safety,* August 1978, pp. 27–8.
13. BOHS Technology Committee Report. Chemical Indicator Tubes for Measurement of the Concentration of Toxic Substances in Air. *Ann. Occup. Hyg.* **16**, 51–62 (1973).
14. K. Leichnitz. *Air Analysis by means of Long-term Detector Tubes.* Drager Review 40, Dec. 1977.
15. B. A. Denenberg, R. S. Kriesel and R. W. Miller. A miniature continuous monitoring system for determination of breathing zone concentrations of toxic substances. Int. Conf. on Environ. Sensing and Assessment, Las Vegas, Sept. 1975.
16. *Methods of Air Sampling and Analysis.* American Public Health Association Intersociety Committee, Washington (1972).
17. L. R. Cook, R. E. Glenn and G. E. Podolak. Monitoring and analysis of personnel exposures to hydrazines at a rocket propellant plant. *Am. Ind. Hyg. Assoc. J.* **40**, 69–74 (1979).
18. H. Moss and E. V. Browett. Determination of tetra-alkyl lead vapour and inorganic lead dust in air. *Analyst,* **91**, 428–38 (1966).
19. US National Institute for Occupational Safety and Health. Dept. of Health Education and Welfare.
20. L. D. White *et al.* A convenient optimised method for the analysis of selected solvent vapours in the industrial atmosphere. *Amer. Ind. Hyg. Assoc. J.* **31**, 225–32 (1970).
21. *NIOSH Manual of Analytical Methods,* 2nd edn. Public 78–175 Department of Health Education and Welfare (NIOSH) Cincinnati (1978).
22. Data from NIOSH Standards Completion Programme published in *Federal Register,* June 1974 plus May 1976 update.
23. R. G. Metcher, R. R. Langner and R. O. Kagel. Criteria for the evaluation of methods for the collection of organic pollutants in air using solid sorbents. *Amer. Ind. Hyg. Assoc. J.* **39**, 349–61 (1978).
24. R. D. Burnett Evaluation of charcoal sampling tubes. *Amer. Ind. Hyg. Assoc. J.* **37**, 37–45 (1976).
25. A. J. Saalwaechter, C. S. McCammon, C. P. Roper and K. S. Carlberg. Performance testing of the NIOSH charcoal tube technique for the determination of air concentrations of organic vapours. *Amer. Ind. Hyg. Assoc. J.* **38**, 476–86 (1977).
26. D. T. Coker. Analysis of Vapour Trapping Tubes used for Personal Monitoring. SIRA Seminar on Monitoring Toxic Gases, Vapours and Particulates, Sheffield, UK, Sept. 1977.
27. R. Jelts. Sampling of nonpolar air contaminants on Poropak porous polymer beads. *Atmos. Environ.* **3**, 587–8 (1969).
28. R. Perry and J. D. Twibell. A time based elution technique for the estimation of specific hydrocarbons in air. *Atmos. Environ.* **7**, 929–37 (1973).
29. J. P. Mieure and M. W. Dietrich. Determination of trace organics in air and water. *J. Chrom. Sci.* **11**, 559–73 (1973).
30. S. A. Myers, H. J. Quinn and W. C. Zook. Determination of VCM at sub-p.p.m. level using a personal monitor. *Amer. Ind. Hyg. Assoc. J.* **36**, 332–7 (1975).

31. J. W. Russell. Analysis of air pollutants using sampling tubes and gas chromatography. *Environ. Sci. Tech.* **9** (13), 1175–8 (1975).
32. D. G. Parkes, C. R. Ganz, A. Polinsky and J. Schulze. A simple gas chromatographic method for the analysis of trace organics in ambient air. *Amer. Ind. Hyg. Assoc. J.* **37**, 165–73 (1976).
33. D. T. Coker. Heat desorption of vapour trapping tubes. *Proc. Analyt. Div. Chem. Soc.* **14** (15), 108–9 (1977).
34. L. W. Severs, R. G. Melcher and M. J. Kocsis. Dynamic U-tube system for solid sorbent air sampling method. *Amer. Ind. Hyg. Assoc. J.* **39**, 321–6 (1978).
35. M. W. Dietrich, L. M. Chapman and J. P. Mieure. Sampling for organic chemicals in workplace atmospheres with porous polymer beads. *Amer. Ind. Hyg. Assoc. J.* **39**, 385–92 (1978).
36. P. Leinster, R. Perry and R. J. Young. Ethylene dibromide in urban air. *Atmos. Environ.* **12**, 2383–7 (1978).
37. D. L. Focrst. A sampling and analytical method for vinyl acetate in air. *Amer. Chem. Soc. 176th Nat. Meet., Miami Beach,* Sept. 1978.
38. D. T. Coker. A simple, sensitive technique for personal and environmental sampling and analysis of lead alkyl vapours in air. *Ann. Occup. Hyg.* **21**, 33–8 (1978).
39. D. A. Ferguson and K. J. Saunders. Quantitative calibration procedures and atmospheric sampling in environmental analysis. *Monitoring of Hazardous Gases in the Working Environment.* Chem. Soc. Int. Conf., City University, London, Dec. 1977.
40. L. H. Nelms, K. D. Reiszner and P. W. West. Personal vinyl chloride monitoring device with permeation technique for sampling. *Analyt. Chem.* **49** (7), 994–8 (1977).
41. M. Evans, M. Molyneux, T. Sharp, A. Bailey and P. Hollingdale-Smith. A practical application of the Porton diffusion sampler for the measurement of time-weighted average exposure to volatile organic substances in air. *Ann. Occup. Hyg.* **20**, 357–63 (1977).
42. A. Bailey and P. Hollingdale-Smith. A personal diffusion sampler for evaluating time weighted exposure to organic gases and vapours. *Ann. Occup. Hyg.* **20**, 345–56 (1977).
43. F. C. Tompkins and R. L. Goldsmith. A new personal dosimeter for the monitoring of industrial pollutants. *Amer. Ind. Hyg. Assoc. J.* **38**, 371–7 (1977).
44. P. W. West and K. D. Reiszner. Field tests for a permeation-type personal monitor for vinyl chloride. *Amer. Ind. Hyg. Assoc. J.* **39**, 645–50 (1978).
45. R. L. Bamberger, G. G. Esposito, B. W. Jacobs, G. E. Podolak and J. F. Mazur. A new personal sampler for organic vapours. *Amer. Ind. Hyg. Assoc. J.* **39**, 701–8 (1978).
46. J. Charlton, K. J. Saunders and R. H. Brown. *Amer. Ind. Hyg. Ass. J.,* in press.
47. P. O. Kane, B. Miller and P. Leinster. Personal communications.
48. A. Ahlmark and S. Forssman. Evaluating trichloroethylene exposures by urinalysis for trichloroacetic acid. *Arch. Ind. Hyg. Occup. Med.* **3**, 386–98 (1951).
49. M. Timar. *Pareszsegugyi Kezlemenyck,* **2**, 36 (1948).
50. L. D. Pagnotto *et al. Amer. Ind. Hyg. Assoc. J.* **22**, 417 (1961).
51. Airborne Lead in Perspective. NAS Report No. ISBN 0-309-01941-9 (1972).
52. *Lead in the Environment and its Significance to Man.* Dept. of Environ. Report. HMSO (1974).
53. R. W. Brandon. Chemically impregnated paper tapes for continuous monitoring of toxic gases. In *Monitoring of Hazardous Gases in the Working Environment.* Chem. Soc. Int. Conf., City University, London, Dec. 1977.
54. R. H. Brown and C. J. Purnell. *J. Chromatog.* **178**, 79 (1979).
55. E. A. Boettner and H. J. Muranko. Animal breath data for estimating the exposure of humans to chlorinated hydrocarbon. *Amer. Ind. Hyg. Assoc. J.* **30**, 437–42 (1969).
56. R. J. Sherwood and F. W. G. Carter. The measurement of occupational exposure to benzene vapour. *Ann. Occup. Hyg.* **13**, 125–46 (1970).
57. R. D. Stewart, C. L. Hake and J. E. Peterson. Use of breath analysis to monitor trichloroethylene exposures. *Arch. Environ. Health,* **29**, 6–13 (1974).

58. US Dept. of Health Educ. and Welfare. *Biologic Standards for the Industrial Worker by Breath Analysis: Trichloroethylene.* NIOSH Publication No. 74–133 (1974).
59. M. Berlin *et al. Biological Index for, and Chromosome Changes in, Benzene Exposure.* Report No. 771018, Lund University, Sweden (1977).
60. K. Enstrom, K. Husman, P. Pfaffli and V. Riikimaki. Evaluation of occupational exposure to xylene by blood, exhaled air and urine analysis. *Scand. J. Work Environ. Health,* **4**, 114–21 (1978).
61. P. O. Droz and J. G. Fernandez. Trichloroethylene exposure: biological monitoring by breath and urine analysis. *Brit. J. Ind. Med.* **35**, 35–42 (1978).
62. J. P. Conkle, B. J. Camp and B. E. Welch. Trace composition of human respiratory gas. *Arch. Environ. Health,* **30**, 290–5 (1975).
63. B. Bott *et al.* British Patent 1 374 575 (1970).
64. J. G. Firth, A. Jones and T. A. Jones. The principles of the detection of flammable atmospheres by catalytic devices. *Comb. Flame,* **21**, 303–11 (1973).
65. M. D. Riihimaki. Kinetics of *m*-xylene in man. Influence of intermittent physical exercise and changing environmental concentrations on kinetics. *Scand. J. Work Environ. Health,* **5**, 232–48 (1979).
66. B. E. Saltzman. *Amer. Ind. Hyg. assoc. J.* **23**, 112 (1962).
67. B. E. Saltzman. *Air Sampling Instruments,* 4th edn. ACGIH (1972).
68. D. T. Coker, A. L. Jones and M. C. Simms. A Practical and Theoretical Assessment of Passive Samplers. *British Occup. Hyg. Soc. Conf.,* April 1981, Nottingham.
69. M. F. Boeniger, D. D. Zaebst, H. R. Ludwig, M. S. Crandall and P. Vongrongsemon. A field comparison of two passive monitors under singular and multiple exposure conditions. *Am. Ind. Hyg. Assoc. Conf.,* May 1979, Chicago.
70. D. T. Coker. Recent developments in personal monitoring for exposure to organic vapours. Int. Environ. Safety Conf., London (Sept. 1980).
71. F. J. Hearl and M. P. Manning. Transient response of diffusion dosimeters. *Amer. Ind. Hyg. Ass. J.* **41**, 778–83 (1980).
72. A. Sato and T. Nakajima. Dose-dependent metabolic interaction between benzene and toluene *in vivo* and *in vitro. Toxicol. Appl. Pharmacol.* **48**, 249–50 (1979).

6 Statistical aspects and air sampling strategies

R. P. Harvey

1 Introduction

Programmes of systematic air sampling in workplaces in which there is air pollution arising through the emission of hazardous gases from plant or processes, and consequent exposure of employees on a routine, daily basis, may be undertaken for one or more of three main reasons. The first of these is to monitor employee exposures to air pollutants and to assess these against criterion levels of exposure which may be either statutory or non-statutory exposure limits based on health risk and other considerations. The second is to monitor the development and continuing effectiveness of engineering provisions for the control of emissions of hazardous gases into the working environment, thereby to indicate when corrective action is required. The third reason is to collect exposure data on certain toxic gases for the purposes of epidemiological studies of exposure-related disease.

The most appropriate strategy to adopt for such programmes and also for occasional occupational hygiene surveys, in terms of the siting, number, duration and frequency of air sampling, would depend upon the objectives and would also depend upon factors such as the kind of gas involved and the type of equipment available for its detection and measurement; the nature of the hazards associated with the gas; the type of process and the circumstances under which emissions of the gas into the working environment occur; and the variability of concentrations of the gas in space and time. The last factor is of particular importance and is the chief concern of this chapter, since it gives rise to the need for statistical approaches to air sampling and decision making.

2 Types and sources of variation

The instantaneous concentrations of an air pollutant arising from plant or processes inside a workroom can vary considerably from point to point within the room and they vary with time during the course of a workshift. An example of temporal fluctuation over a fairly short period of time at a fixed point in a room is shown in Fig. 6.1.

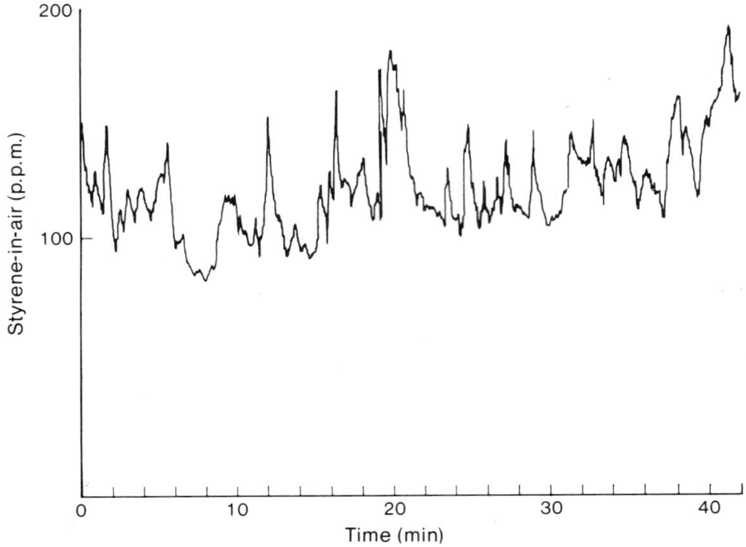

Fig. 6.1 Part of a continuous recording of styrene-in-air concentrations measured by a Miran infra-red gas analyser at a point 1.5 m from a GRP boat mould during laminating work.

In this case the pollutant is styrene vapour emitted from the surface of a mould during the manufacture of a glass reinforced plastic boat.

In addition to such within-shift variation of pollutant concentrations, the average concentration for either a single point in a workroom or the average for a number of points, varies from shift to shift. An example of variation in daily and weekly average concentrations of a pollutant gas in a chemical plant room is shown in Fig. 6.2. These averages were calculated from data obtained by an automatic, multi-point sequential air monitoring system. Similar examples have been given by Peterson, Hoyle and Schneider[1] and Dharmarajan, Weil and Self.[2]

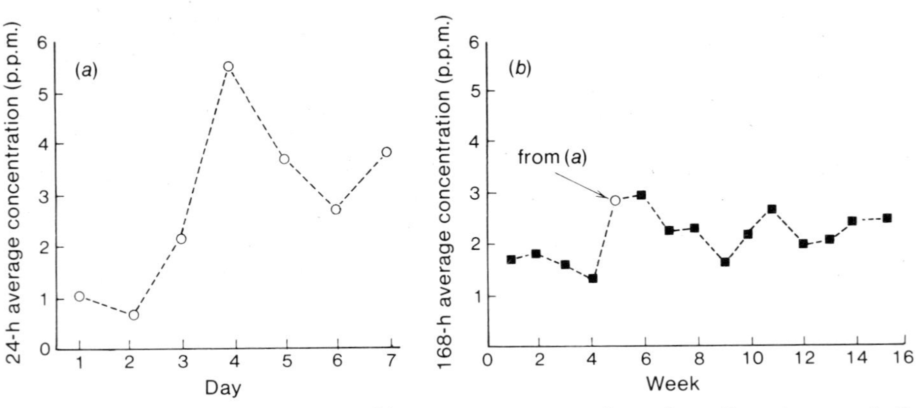

Fig. 6.2 Variation of daily and weekly average concentrations of a pollutant vapour in the air of a chemical plant room. (*a*) Daily averages over a period of one week; (*b*) weekly averages over a period of 15 weeks.

Both the within-shift and between-shift variations (time based) that have been described can be due to combinations of factors such as those given below, acting randomly or systematically:

(*a*) Variation in the number of sources of emission of the pollutant gas for a variety of reasons (for example: expansion or contraction of production operations and processes; discontinuities in production processes; changes in work patterns; intermittent breakdowns; spillages and leaks; improvement or deterioration in plant maintenance and contaminant control).

(*b*) Variation in the rate of emission of the pollutant from a given source (due for example to: variation in the composition of materials used in the processes involved; changes in rate of production; changes in the effectiveness of local exhaust ventilation control; variation in room temperature affecting evaporation rates; etc.).

(*c*) Variation in the dispersion of a pollutant from its source of emission, and in dilution, due to aerodynamic factors. These include: (i) brief fluctuations in the speed and direction of air currents in the workroom due to the movement of equipment and people, external wind effects, etc. and (ii) changes in ventilation rate and in the general aerodynamics of the room due to factors such as changes in external wind conditions, seasonal changes in temperature and in the number of natural ventilation openings and changes in mechanical ventilation provisions.

The concentrations of a pollutant in the breathing zones of employees in a workroom or plantroom vary within and between shifts partly owing to the factors which have just been mentioned and partly owing to additional factors which contribute to variation in individual employees' exposures and also to differences between individuals. These factors include the employees' own changing positions in relation to the sources of emission and differences among themselves in the nature of their jobs and in their work habits.

An example of variation in breathing zone concentrations of a pollutant for one person, measured on a continuous basis over an 8-h shift, is shown in Fig. 6.3. The person was engaged in making glass reinforced plastic boat mouldings, which involved exposure to styrene vapour. His 8-h time-weighted average exposure level was 68.5 p.p.m. It can be seen from Fig. 6.3 that there were marked changes in the level of exposure over the 8-h shift, corresponding to changes in the nature of the work done. There are many jobs in industry for which within-shift exposures to air pollutants are likely to follow changing patterns of this kind.

During each of the stages of work in which the person was close to the source of the styrene (e.g. during lamination of the hull moulding) there were apparently random fluctuations in concentration superimposed over a gradual rise in average level of styrene-in-air as the stage progressed. The latter effect was due to a combination of the increasing area of styrene emitting surface and poor ventilation in the work zone. This effect is emphasised in Fig. 6.4 which compares the fluctuating concentrations of styrene-in-air experienced during the final stage of hull lamination under poor ventilation conditions (Fig. 6.4 *a* is an enlarged part of the curve shown in Fig. 6.3) with those experienced when the same work was done with a local exhaust ventilation system installed. Further reference to these curves will be made later.

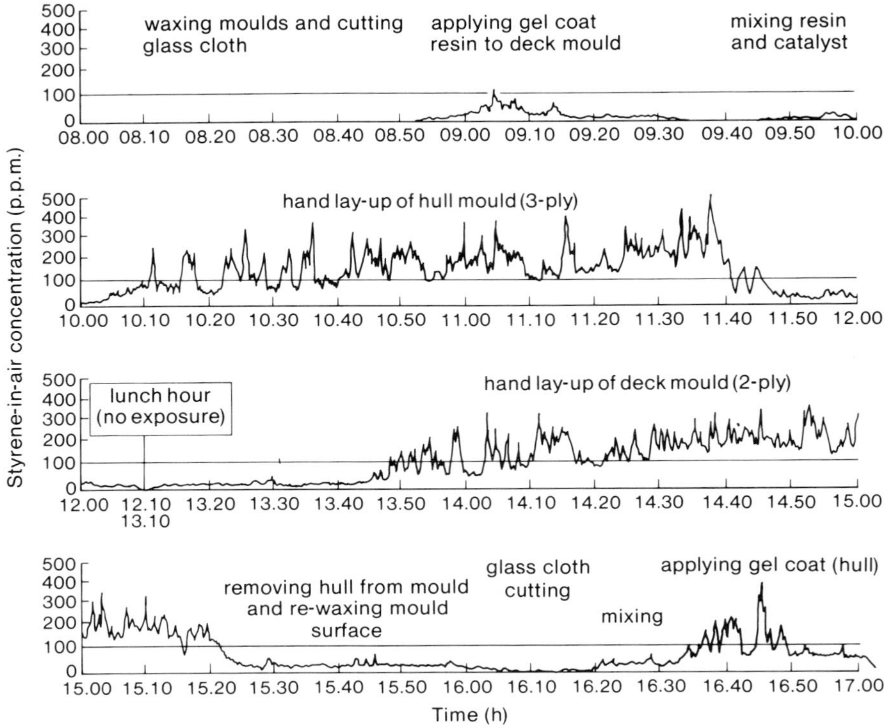

Fig. 6.3 Continuous recording of styrene-in-air concentrations in the breathing zone of a person doing glass reinforced plastic moulding work. His 8-h time weighted average exposure was 68.5 p.p.m.

An example of between-shift differences in 8-h average breathing zone concentrations of a pollutant experienced by employees doing similar jobs at a particular industrial process is shown in Fig. 6.5. In this example the pollutant was a mixture of dust and fume rather than a gas or vapour, but similar kinds of variation can be expected for gases and vapours at many chemical processes. For each of the four weeks that are depicted (each separated by one- or two-month intervals) the shift average exposures for the two employees were measured over four consecutive days. There were random differences between the consecutive days in each week and between the two employees. There were also considerable differences between each of the four weeks.

It is often the case that breathing zone measurements made during a particular work activity indicate a personal exposure to higher instantaneous or short term average concentrations of a pollutant than those found in the workroom air generally by static or fixed point air sampling over the same period. This occurs when the person is working close to the source of emission in a ventilated room. However, for 8-h or longer averaging periods, personal exposures are not always higher than general workroom or work zone concentrations. The position might be reversed under circumstances where employees enter polluted air zones for only part of the

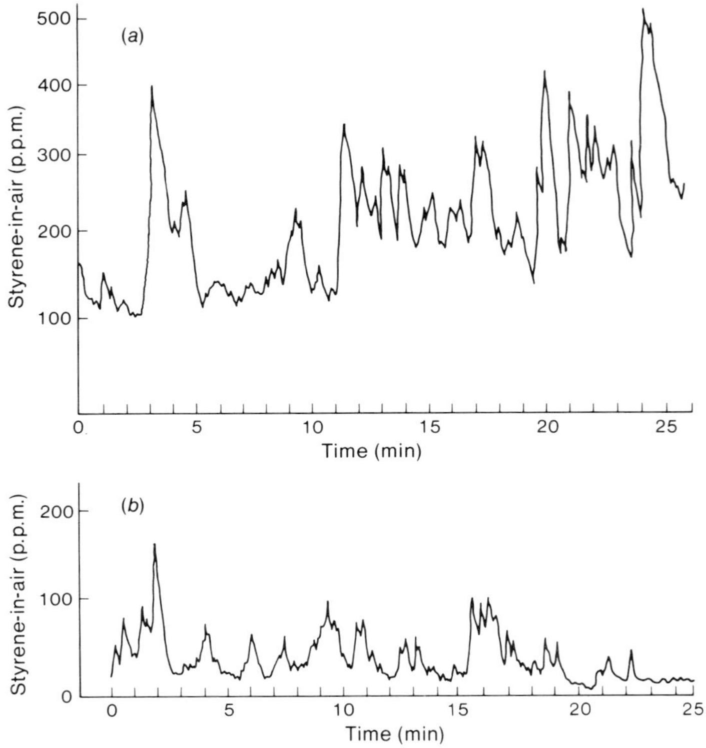

Fig. 6.4 Continuous record of styrene-in-air concentrations in the breathing zone of a person hand laminating the third ply of a 4.5 m long glass reinforced plastic sailing boat hull: (*a*) under poor general ventilation conditions; (*b*) under local exhaust ventilation conditions.

workshift, as for example might occur with certain types of work in the chemical industry.

There are important implications here for air sampling strategies. The fact that pollutant concentrations measured in employees' breathing zones can differ substantially from the results of fixed point measurements indicates that in any air sampling programme designed either to assess exposures against a limit value which is a hygiene standard or to collect data for epidemiological purposes, the emphasis should be on the use of personal sampling techniques. The use of fixed point sampling for these purposes would be likely to give very misleading results in many circumstances. However, the latter would be expected to be the more appropriate approach for monitoring the performance of provisions for the control of emissions at certain types of process.

3 Theoretical frequency distributions

It may be of considerable value for the purposes of health risk assessment, for decision-making on various matters relating to the management and control of a

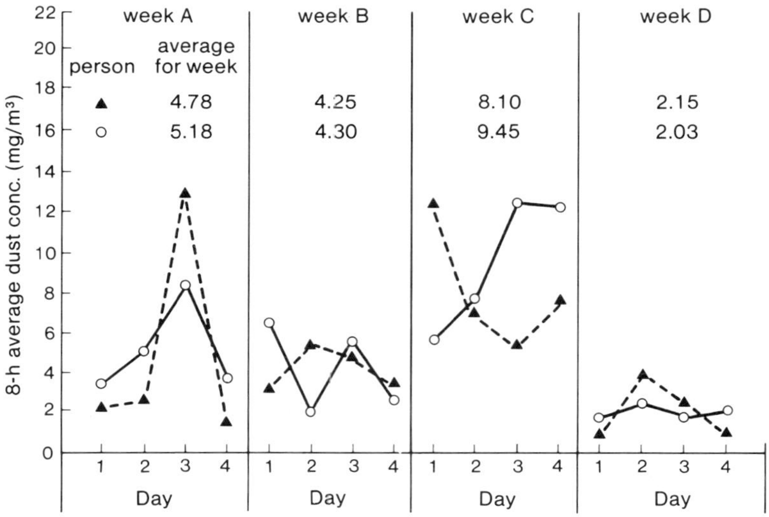

Fig. 6.5 Variation in full shift (8 h) time weighted average exposures of two workers to dust and fume in a particular job.

working environment, and also for the purposes of special studies of exposure effects, to be able to use a mathematical description of the variability of time averaged pollutant concentrations that are observed over a specified assessment period at either fixed locations in the environment or in the breathing zones of the members of a well defined group of exposed employees. The assessment time in question might be one shift, one week, one year or more. It should be borne in mind, of course, that the variability with which we are concerned here, arises only when the averaging time of pollutant concentration measurements is short in relation to the assessment time. If the variability of time averaged concentrations could, in fact, be described by the probability density function for a theoretical frequency distribution, then it would be possible to make statements either about the probability of occurrence of any specific concentration value, or about the probability of observing concentrations above or below some criterion value.

The frequency distributions of the results of within-shift measurements of either instantaneous concentrations or short-term average concentrations of air pollutants typically have a positive skew and so too do the frequency distributions of shift averages. Thus the probability density functions that have attracted interest in the study of air pollutant variability are those for non-symmetrical distributions such as the exponential, gamma and Weibull distributions[3–6] and, more especially, the two-parameter and three-parameter lognormal distributions.[7] The properties of the latter two probability models and their goodness of fit to air sampling data from various sources are described below. The properties of the normal probability model are also considered, partly because of its obvious relation with the lognormal model and partly because it describes the possible distribution (the sampling distribution) of repeated estimates, $\bar{x}$, of the arithmetic mean pollutant concentration for a given assessment time, T, each obtained from n random short term pollutant-in-air measurements of duration t hours (this may not be true, generally, when n is small). It may also describe the distribution of overall random measuring errors. The normal distribution is thus used as a basis for certain statistical decision procedures.

Two concepts which may need to be explained at this point are those of a statistical population and a statistical sample (not to be confused with the term 'air sample'). A statistical population is an infinite or a finite (but large) set of elements, each having in common a variable property which is of particular interest and can be measured. If the property of interest were a random variable and were measured on all the elements of the population, or on a very large number of them, then sorting of the measurements into ranked groupings would reveal the population distribution of the property, which would have certain characteristics. Such characteristics or parameters include the mean and variance.

A statistical sample is a relatively small number (a subset) of the elements drawn from the population. For practical and economic reasons, measurements of the property of interest on the elements of the subset may be used to determine estimates of, and to make inferences about, the parameters of the whole population. For this purpose the elements comprising the sample subset would generally have to be drawn at random from the total set by simple random sampling, or by stratified random sampling if necessary, but under some circumstances (e.g. where there is no likelihood of cyclic variation in the value of the property with time) they may be drawn by systematic sampling.[8,9]

In the present context a population of interest might be, for example, the set of 8-h time averaged pollutant exposure levels $x_1, x_2, \ldots x_N$ experienced by a group of twenty employees over an assessment period of 2 yr, with $N = 4800$. A statistical sample from that population might be a set of $n = 60$ personal measurements of 8-h averaged concentrations taken either at random or systematic intervals over the 2-yr period to provide an estimate of the population average.

3.1 The normal distribution

(*a*) *Properties of the population distribution*

The probability density function for the normal distribution of a random variate X which is a property of an infinite population of elements is given by

$$f(x) = \frac{1}{\sigma\sqrt{2\pi}} \exp\left\{-\frac{1}{2\sigma^2}(x-\mu)^2\right\} \tag{6.1}$$

for $-\infty < x < \infty$, where μ = population mean; σ^2 = population variance.

This function will give the probability of obtaining a particular value x of the random variate X. It is a necessary property of all probability density functions that the area under the curve of the function is equal to one, i.e.

$$\int_{-\infty}^{\infty} f(x)\,dx = 1 \tag{6.2}$$

If it is required to know the probability of obtaining a value of X which is less than x_j, then this is given by integration of the function between the limits $-\infty$ and x_j, i.e.

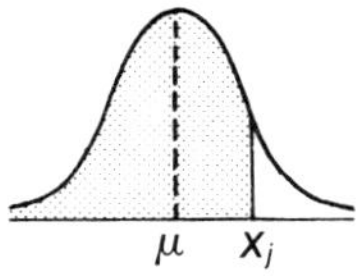

$$P(x \leq x_j) = \int_{-\infty}^{x_j} f(x)\,\mathrm{d}x \tag{6.3}$$

and similarly,

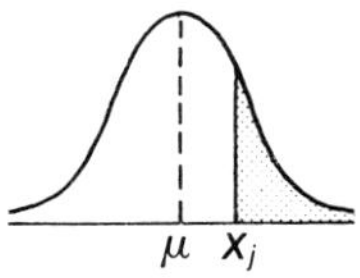

$$P(x \geq x_j) = 1 - \int_{-\infty}^{x_j} f(x)\,\mathrm{d}x \tag{6.4}$$

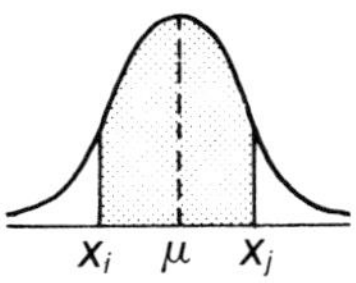

$$P(x_i \leq x \leq x_j = \int_{x_i}^{x_j} f(x)\,\mathrm{d}x \tag{6.5}$$

If the normal distribution is standardised, areas under the curve (probabilities) for given intervals of the variate X, can be tabulated and applied to any normal distribution whatever the values of the mean and variance. This is done by forming a new random variable, Z, which is normally distributed with a mean of zero and a variance of one, where

$$Z = \frac{x - \mu}{\sigma} \tag{6.6}$$

The probability density function for the standardised normal distribution is thus

$$f(Z) = \frac{1}{\sqrt{2\pi}} \exp\left\{-\frac{Z^2}{2}\right\} \tag{6.7}$$

Areas under one tail of this distribution are usually given in statistical tables.[10] These are calculated from the distribution function, sometimes denoted by $\Phi(Z)$, which is expressed as

$$P(Z \geq Z_\alpha) = \alpha = \int_{Z_\alpha}^{\infty} \frac{1}{\sqrt{2\pi}} \exp\left\{-\frac{Z^2}{2}\right\} \tag{6.8}$$

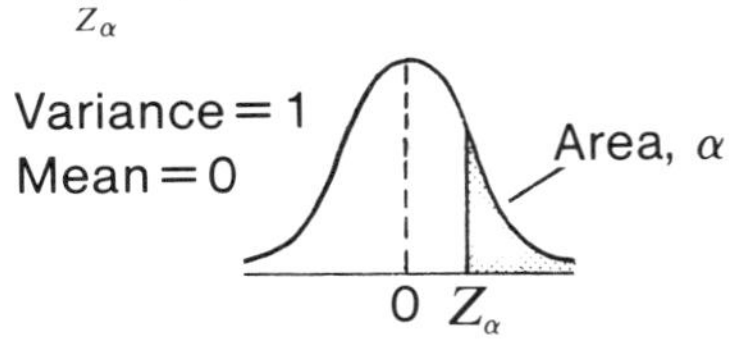

Thus Z_α is the value of the standard normal variate which has a probability α of being exceeded. Some values of Z for given values of α are shown in Table 6.1.

Percentage points or quantiles of the population distribution of X can be determined from the standardised normal distribution using

$$\text{for 50\% quantile and below,} \quad Q_{100\alpha} = \mu - Z_\alpha \sigma \tag{6.9}$$

$$\text{for 50\% quantile and above,} \quad Q_{100(1-\alpha)} = \mu + Z_\alpha \sigma \tag{6.10}$$

Thus to determine the 95% quantile, for example, the value $Z_\alpha = 1.645$ (corresponding to $\alpha = 0.05$, see Table 6.1) is substituted into Equation 6.10. This would then give that value of X below which 95% of all values lie.

(b) Estimation of parameters μ and σ^2

Now, it is often the case that the true mean and variance (μ and σ^2 respectively) of the population distribution are not known and that estimates of these parameters are obtained from relatively small samples of size n, as mentioned earlier. These estimates are given by:

$$\text{sample mean,} \quad \bar{x} = \frac{1}{n}\sum_i^n x_i \tag{6.11}$$

$$\text{sample variance,} \quad S^2 = \frac{\Sigma(x_i - \bar{x})^2}{n-1} = \frac{\Sigma x_i^2 - 1/n(\Sigma x_i)^2}{n-1} \tag{6.12}$$

for $i = 1, 2, \ldots n$. Note that as $n \to \infty$, so $\bar{x} \to \mu$ and $S^2 \to \sigma^2$.

(c) Confidence intervals for population parameters

The precision of these estimates depends upon the manner in which the sample was drawn from the population, the size, n, of the sample, and the characteristics of the population distribution. A measure of the precision of the estimators $\bar{x}$ and S^2 can be obtained by considering their respective sampling distributions. For repeated samples each having n elements drawn at random from a normal population, it can be shown from purely mathematical considerations and also from sampling experiments, that the distribution of the sample means $\bar{x}$ (referred to as the sampling distribution of the random variable $\bar{X}$), is also normal regardless of the size of n, and that the variance is σ^2/n. The square root of this expression, $\sigma/\sqrt{n}$, is called the standard error of the mean and is sometimes denoted $\sigma_{\bar{x}}$.

With regard to the sampling distribution of the estimator S^2, it can be shown both

Table 6.1 Areas under the tail of the standardised normal distribution

Single tail area (α)	Z
0.05	1.645
0.025	1.960
0.01	2.326
0.001	3.090

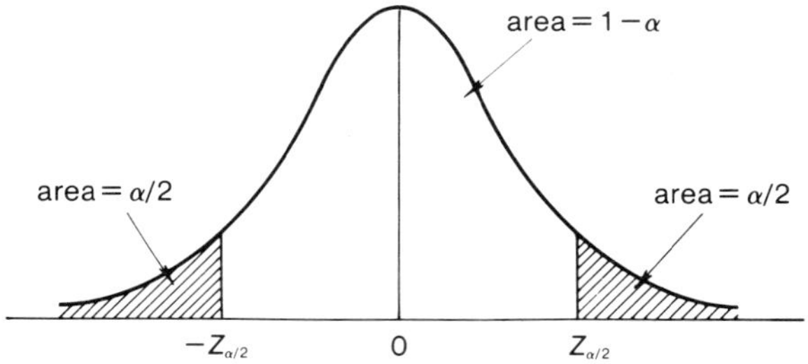

Fig. 6.6 Illustration of the confidence interval for μ.

empirically and from purely mathematical considerations that if the parent population distribution is normal, the statistic

$$\frac{(n-1)\,S^2}{\sigma^2}$$

has a chi-square distribution with $n-1$ degrees of freedom.

Under the central limit theorem,[5] the above statements about sampling distributions of $\overline{X}$ also hold when the parent population distribution is asymmetrical, that is skew. This theorem states amongst other things that the sampling distribution of the means of random samples will be approximately normal regardless of the form of the population distribution of X, provided that the sample size, n, is sufficiently large. The more the population distribution differs from a normal distribution, the larger the sample is likely to have to be for the theorem to hold. An empirical vertification of the theorem using data derived from the styrene exposure curve in Fig. 6.3, is given in a later section.

For both of these cases, that is for a sample drawn from either a normal or a non-normal parent population, the sampling distribution used for estimating the precision of the sample mean and for formulating decision rules is that of the standardised mean (provided that n is fairly large):

$$Z = \frac{\overline{x}-\mu}{\sigma/n} \tag{6.13}$$

The limiting distribution of Z (as $n \rightarrow \infty$) is in fact the standardised normal distribution, which as already described, has a mean of zero and variance of one. This can be used to make probability statements about a sample mean, $\overline{x}$, or to specify the interval within which the true population mean μ would be expected to fall with a given degree of confidence.

By considering both sides of this distribution, bearing in mind its symmetry (Fig. 6.6) and letting $Z_{\alpha/2}$ be such that integral of Equation 6.7 from $Z_{\alpha/2}$ to ∞ equals $\alpha/2$, it can be asserted with a probability of $(1-\alpha)$ that the standardised mean Z will assume a value between $-Z_{\alpha/2}$ and $Z_{\alpha/2}$, that is in the interval

$$-Z_{\alpha/2} < \frac{\overline{x}-\mu}{\sigma/\sqrt{n}} < Z_{\alpha/2} \tag{6.14}$$

This double inequality is algebraically equivalent to

$$\bar{x} - Z_{\alpha/2}\,\sigma/\sqrt{n} < \mu < \bar{x} + Z_{\alpha/2}\,\sigma/\sqrt{n} \qquad (6.15)$$

which gives the $(1-\alpha)$ or $100(1-\alpha)\%$ confidence interval for μ, for the case where the true variance, σ^2, of the population is known. The left side of the inequality is called the lower confidence limit, LCL, and the right side is called the upper confidence limit, UCL. From the values given in Table 6.1, it can be seen that the $100(1-0.05)\%$ or 95% confidence interval (bearing in mind that for the two tailed condition we are concerned with $Z_{\alpha/2}$ where $\alpha/2 = 0.025$) is given by

$$\bar{x} - 1.96\,\sigma/\sqrt{n} < \mu < \bar{x} + 1.96\,\sigma/\sqrt{n} \qquad (6.16)$$

If only one confidence limit and hence one tail of the standardised normal sampling distribution of Z is of interest, that is if it is required to know either the limit above which there is a probability of $100(1-\alpha)\%$ that μ will fall, or the limit below which there is a $100(1-\alpha)\%$ probability that μ will fall, then such single tailed confidence limits are given respectively by

$$\mathrm{LCL}_{100(1-\alpha)\%} = \bar{x} - Z_{\alpha}\,\sigma/\sqrt{n} \qquad (6.17)$$

and $$\mathrm{UCL}_{100(1-\alpha)\%} = \bar{x} + Z_{\alpha}\,\sigma/\sqrt{n} \qquad (6.18)$$

From Table 6.1, the single tailed 95% confidence limits would be given by

$$\mathrm{LCL}_{95} = \bar{x} - 1.645\,\sigma/\sqrt{n} \qquad (6.19)$$

and $$\mathrm{UCL}_{95} = \bar{x} + 1.645\,\sigma/\sqrt{n} \qquad (6.20)$$

If the population distribution variance, σ^2, is not known then the estimate S^2 obtained from the sample subset, or rather its square root S (standard deviation) is used in Equation 6.15 to obtain an approximate confidence interval provided that the number of elements, n, is greater than about 30. Similarly, the estimate S is used in Equations 6.17 and 6.18 to estimate single tail confidence limits.

Where the population variance σ^2 is unknown and the number of elements, n, in the sample subset (the sample size) is small, the standardised normal distribution should not be used. In this case the t-distribution is used to specify a confidence interval for μ, but only if the parent population or the sampling distribution of $\bar{X}$ is known, or can be safely assumed, to be normal. The two tailed $100(1-\alpha)\%$ confidence interval is then given by

$$\bar{x} - t_{\alpha/2,\nu}\,S/\sqrt{n} < \mu < \bar{x} + t_{\alpha/2,\nu}\,S/\sqrt{n} \qquad (6.21)$$

In this expression the term $S/\sqrt{n}$ is an estimate of the standard error of the mean, sometimes denoted as $S_{\bar{x}}$. The value of $t_{\alpha/2,\nu}$ can be obtained by reference to a table of the probability points of the t-distribution[10] and corresponds to a given area (probability) under each tail of the curve and to $\nu = n-1$ degrees of freedom. Some values of the t-distribution for probabilities of $\alpha = 0.05$ and $\alpha = 0.025$ are given in Table 6.2. It should be noted that as the sample size, n, increases, so the value of t for a given probability, α, approaches that of Z.

The probabilities in Table 6.2 refer to only one tail of the t-distribution so that if, for example we have a sample size of $n = 11$, then a value of $t_{0.025,10} = 2.23$ would

Table 6.2 Percentage points of the t-distribution for v degrees of freedom

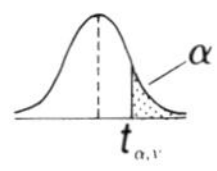

Degrees of freedom v	$\alpha = 0.05$	$\alpha = 0.025$
1	6.31	12.71
2	2.92	4.30
5	2.01	2.57
10	1.81	2.23
20	1.73	2.09
30	1.70	2.04
∞	1.65	1.96

have to be used in Equation 6.21 to estimate the $100(1-0.05)\%$ or 95% confidence interval for μ, bearing in mind the symmetry of the distribution. If the single tail upper or lower 95% confidence limit were required for a sample of size $n = 11$, then $t_{0.05,10} = 1.81$ would be used as follows,

$$\mathrm{LCL}_{95} = \bar{x} - 1.81\, S/\sqrt{n}$$

or
$$\mathrm{UCL}_{95} = \bar{x} + 1.81\, S/\sqrt{n}$$

So far, only infinite populations have been considered. If we have a finite population of known size, N, from which a sample of size n (having mean $\bar{x}$, and standard deviation S) is drawn, then a correction has to be applied to the standard error of the mean. This is particularly important when the sample size is large compared with N, that is when $n/N > 0.05$. In the case of a finite population, the standard error of the mean is estimated by

$$S_{\bar{x}} = \frac{S}{\sqrt{n}}\sqrt{\frac{N-n}{N-1}} \tag{6.22}$$

Expression 6.21 for the $100(1-\alpha)\%$ confidence interval for the mean is thus modified to:

$$\bar{x} - t_{\alpha/2,v}\frac{S}{\sqrt{n}}\sqrt{\frac{N-n}{N-1}} < \mu < \bar{x} + t_{\alpha/2,v}\frac{S}{\sqrt{n}}\sqrt{\frac{N-n}{N-1}} \tag{6.23}$$

On the basis of the earlier observation about the sampling distribution of the statistic $S^2(n-1)1/\sigma^2$, the $100(1-\alpha)\%$ confidence interval for the infinite population variance, σ^2, can be expressed in terms of the estimate S^2 obtained from a sample of size n, provided that the sample is drawn from a normal population, as follows,

$$\frac{(n-1)\,S^2}{\chi^2_{\alpha/2,v}} < \sigma^2 < \frac{(n-1)\,S^2}{\chi^2_{1-\alpha/2,v}} \tag{6.24}$$

where $\chi^2_{\alpha/2,v}$ and $\chi^2_{1-\alpha/2,v}$ are values of chi-square, given by tables of the chi-square distribution,[10] which are associated with areas under the curve (probabilities) of $\alpha/2$ and $1-\alpha/2$ and with $\nu=n-1$ degrees of freedom.

3.2 The two-parameter lognormal distribution

(a) Properties of the population distribution

If the behaviour of a random variate X (which is some property of a population of elements) is such that the transformed variate $Y=\ln X$ has a normal population distribution with mean μ and variance σ^2, then the probability density function for the distribution of X itself can be expressed in terms of these same two parameters as

$$f(x)=\frac{1}{x\sigma\sqrt{2\pi}}\exp\left\{-\frac{1}{2\sigma^2}(\ln x-\mu)^2\right\} \tag{6.25}$$

for $0<x<\infty$.

The mean and variance of the lognormal population distribution which Equation 6.25 defines, can be expressed in terms of the mean μ and variance σ^2 of the population distribution of Y, as follows:

$$\text{population mean,} \quad \theta=\exp\{\mu+\tfrac{1}{2}\sigma^2\} \tag{6.26}$$

$$\text{population variance,} \quad \beta^2=[\exp\{2\mu+\sigma^2\}][\exp\{\sigma^2\}-1] \tag{6.27}$$

These expressions are derived from the first moment about the origin and the second moment about the mean, respectively.

Two other population parameters of the lognormal distribution of X can also be used to specify that distribution completely. These parameters are the median or geometric mean, μ_g, and the geometric standard deviation, σ_g, which are given by

$$\mu_g=\exp(\mu) \tag{6.28}$$

$$\sigma_g=\exp(\sigma) \tag{6.29}$$

From these definitions it follows that alternative expressions for the population mean θ can be used, as follows:

$$\theta=\exp\{\mu\}\exp\{\tfrac{1}{2}\sigma^2\}=\mu_g\exp\{\tfrac{1}{2}(\ln\sigma_g)^2\} \tag{6.30}$$

or

$$\theta=\exp\{\ln\mu_g+\tfrac{1}{2}(\ln\sigma_g)^2\} \tag{6.31}$$

From Equations 6.9 and 6.10 it follows that quantiles of the log-normal distribution are given by,

$$\text{for 50\% quantile and below,} \quad Q'_{100\alpha}=\exp\{\mu-Z_\alpha\sigma\} \tag{6.32}$$

$$\text{for 50\% quantile and above,} \quad Q'_{100(1-\alpha)}=\exp\{\mu+Z_\alpha\sigma\} \tag{6.33}$$

The 95% quantile, which is often of interest in air pollutant studies, is obtained by substituting the appropriate value of Z_α from Table 6.1 into Equation 6.33 to give

$$Q'_{95}=\exp\{\mu+1.645\,\sigma\} \tag{6.34}$$

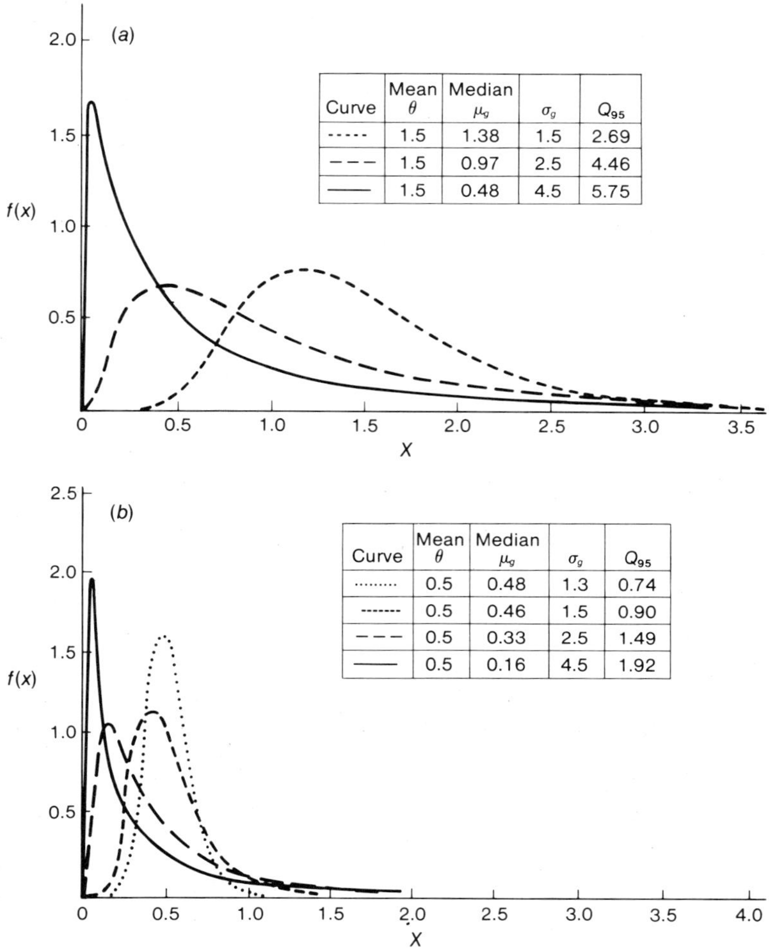

(a)

Curve	Mean θ	Median μ_g	σ_g	Q_{95}
- - - - -	1.5	1.38	1.5	2.69
— — —	1.5	0.97	2.5	4.46
———	1.5	0.48	4.5	5.75

(b)

Curve	Mean θ	Median μ_g	σ_g	Q_{95}
········	0.5	0.48	1.3	0.74
-------	0.5	0.46	1.5	0.90
— — —	0.5	0.33	2.5	1.49
———	0.5	0.16	4.5	1.92

Fig. 6.7 Probability curves for lognormal distributions with (*a*) $\theta = 1.5$ and (*b*) $\theta = 0.5$ and with different geometric standard deviations.

Three alternative expressions for Q'_{95} by virtue of Equations 6.28 to 6.31 are,

$$Q'_{95} = \exp\{\ln \mu_g + 1.645 \ln \sigma_g\} \tag{6.35}$$

$$Q'_{95} = \mu_g\, \sigma_g^{1.645} \tag{6.36}$$

$$Q'_{95} = \theta \exp\{1.645 \ln \sigma_g - \tfrac{1}{2}(\ln \sigma_g)^2\} \tag{6.37}$$

Lognormal distribution curves for populations having different geometric standard deviations and having means of $\theta = 1.5$ and $\theta = 0.5$ are shown in Fig. 6.7(*a*) and (*b*) respectively. The values of the 95% quantiles, calculated from Equation 6.37 are indicated. These curves demonstrate that the lognormal probability density function is capable of describing a wide variety of shapes of frequency distribution. It may be noted that as the geometric standard deviation decreases and approaches a value of 1.0, so the degree of skewness diminishes and the shape of the distribution changes from an extreme positive skew to approach the bell shape of the normal distribution.

(*b*) *Estimation of population parameters*

The parameters of the normal population distribution of Y and of the lognormal population distribution of X are usually estimated, at least initially, from relatively small sample subsets of n elements. In the case of the distribution of Y, the method of maximum likelihood gives the following minimum variance unbiased estimators of μ and σ^2 respectively:

$$\text{sample mean,} \quad \bar{y} = \frac{1}{n}\Sigma y_i \tag{6.38}$$

$$\text{sample variance,}\ S_y^2 = \frac{1}{n-1}\left[\Sigma y_i^2 - \frac{1}{n}\left(\Sigma y_i\right)^2\right] \tag{6.39}$$

where $y = \ln x$ and $i = 1, 2, \ldots n$. As $n \to \infty$, so $\bar{y} \to \mu$ and $S_y^2 \to \sigma^2$.

In the case of the lognormal distribution of X, the estimators of the parameters of interest in workroom and external air pollution studies, that is the mean θ, the geometric mean μ_g and the geometric standard deviation σ_g, are given as follows:

$$\text{sample geometric mean,}\ M_g = \exp\{\bar{y}\} \tag{6.40}$$

$$\text{sample geometric standard deviation,}\ S_g = \exp\{S_y\} \tag{6.41}$$

and for the first moment of the distribution (the mean, θ) three possible estimators are,

$$\text{(i)}\ M_a' = 1/n\,\Sigma x_i = \bar{x} \tag{6.42}$$

$$\text{(ii)}\ M_a'' = M_g\,\psi(t) \tag{6.43}$$

$$\text{where}\ \psi(t) = \frac{1+(n-1)t}{n} + \frac{(n-1)^3 t^2}{n^2(n+1)2!} + \frac{(n-1)^5 t^3}{n^3(n+1)(n+3)3!} + \ldots \tag{6.44}$$

and where $t = \frac{1}{2} S_y^2$

$$\text{(iii)}\ M_a''' = \exp\{\bar{y} + \tfrac{1}{2} S_y^2\} = M_g \exp\{\tfrac{1}{2}(\ln S_g)^2\} \tag{6.45}$$

The first of these estimators of θ, that is the simple arithmetic mean M_a' (or $\bar{x}$) does not, theoretically, have minimum variance. This has been verified empirically by Esmen and Hammad[11] using air pollutant data. These authors also showed that $\bar{x}$ tended to underestimate θ, particularly if the sample size was small ($n < 20$), and it followed a downward trend in efficiency for distributions having geometric standard deviations of greater than 1.5. Aitchison and Brown[7] have shown the latter effect to be predicted theoretically.

The second estimator, M_a'', which is due to Finney,[12] is the best in that theory

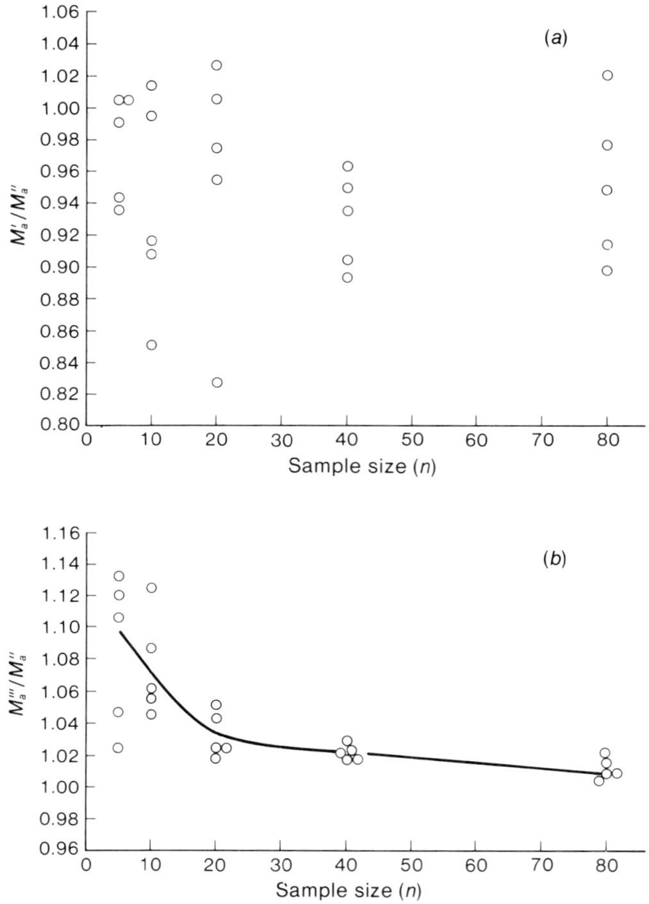

Fig. 6.8 Comparison of estimates M'_a and M'''_a with the best minimum variance unbiased estimator, M''_a, by ratios: (*a*) ratio of M'_a to M''_a; (*b*) ratio of M'''_a to M''_a. The samples of different sizes were drawn at random (with replacement) from a finite lognormal distribution.

predicts it to be a minimum variance unbiased estimator. It may appear to be somewhat tedious to use, but tables of the function $\psi(t)$ can be prepared with the aid of a computer. The empirical data in Fig. 6.8, which compares the ratios of M'_a to M''_a *and of* M'''_a to M''_a, indicate that the third estimator M'''_a provides estimates which are close to those of M''_a for sample sizes of $n > 20$. These data were obtained from a sampling experiment in which estimates of θ were made for each of five random samples of size $n = 5$, 10, 20, 40 and 80 in turn, drawn from an artificial lognormal distribution of 1092 time-averaged concentrations having a mean $\theta = 5.47$, median or geometric mean, $\mu_g = 3.32$ and a geometric standard deviation, $\sigma_g = 2.72$. One important observation to make from Fig. 6.8(*a*) is that it confirms the tendency for M'_a (or $\bar{x}$) to underestimate the mean θ of a lognormal distribution. This appears to be true even for large sample sizes. Indeed, the arithmetic mean for a total of 1092 concentration values was only 4.98, whereas the mean θ was 5.47 as already noted above.

The best estimator of the population variance, β^2, of the lognormal distribution of X is

$$S_a^2 = M_g^2 \{\psi(t_1) - \psi(t_2)\} \tag{6.46}$$

where the functions $\psi(t_1)$ and $\psi(t_2)$ are defined by Equation 6.44 and where

$$t_1 = 2S_y^2 \quad \text{and} \quad t_2 = \left(\frac{n-2}{n-1}\right) S_y^2$$

An alternative method of estimating β^2 is to substitute estimates of μ and σ^2 into Equation 6.27.

(c) Confidence intervals for population parameters

The $100(1-\alpha)\%$ confidence intervals for the mean μ of the distribution of Y and for the geometric mean, μ_g, of the distribution of X, given that estimates $\bar{y}$ and S_y are obtained from a sample of size n drawn from an infinite population (which is normal for Y and lognormal for X), are respectively

$$\bar{y} - t_{\alpha/2,\nu} \frac{S_y}{\sqrt{n}} < \mu < \bar{y} + t_{\alpha/2,\nu} \frac{S_y}{\sqrt{n}} \tag{6.47}$$

and

$$\exp\left\{\bar{y} - t_{\alpha/2,\nu} \frac{S_y}{\sqrt{n}}\right\} < \mu_g < \exp\left\{\bar{y} + t_{\alpha/2,\nu} \frac{S_y}{\sqrt{n}}\right\} \tag{6.48}$$

or

$$M_g \exp\left\{-t_{\alpha/2,\nu} \frac{\ln S_g}{\sqrt{n}}\right\} < \mu_g < M_g \exp\left\{t_{\alpha/2,\nu} \frac{\ln S_g}{\sqrt{n}}\right\} \tag{6.49}$$

The $100(1-\alpha)\%$ confidence intervals for the standard deviation σ of the Y distribution and for the geometric standard deviation σ_g of the X distribution, for any sample size, are respectively,

$$\left\{\frac{(n-1)\,S_y^2}{X^2_{\alpha/2,\nu}}\right\}^{0.5} < \sigma < \left\{\frac{(n-1)\,S_y^2}{X^2_{1-\alpha/2,\nu}}\right\}^{0.5} \tag{6.50}$$

and

$$\exp\left\{\left[\frac{(n-1)\,S_y^2}{X^2_{\alpha/2,\nu}}\right]^{0.5}\right\} < \sigma_g < \exp\left\{\left[\frac{(n-1)\,S_y^2}{X^2_{1-\alpha/2,\nu}}\right]^{0.5}\right\} \tag{6.51}$$

There is no way of obtaining exact confidence intervals for the mean θ and variance β^2 of the lognormal distribution, on the basis of theory. If the parent population has a finite size N, then following the argument which led to Equation 6.23 in Section 3.1 (*c*), the confidence intervals given by Equations 6.47 and 6.49 are modified to

$$\bar{y}-t_{\alpha/2,v}\frac{S_y}{\sqrt{n}}\sqrt{\frac{N-n}{N-1}}<\mu<\bar{y}+t_{\alpha/2,v}\frac{S_y}{\sqrt{n}}\sqrt{\frac{N-n}{N-1}} \tag{6.52}$$

and

$$M_g \exp\left\{-t_{\alpha/2,v}\frac{\ln S_g}{\sqrt{n}}\sqrt{\frac{N-n}{N-1}}\right\}<\mu_g<M_g \exp\left\{t_{\alpha/2,v}\frac{\ln S_g}{\sqrt{n}}\sqrt{\frac{N-n}{N-1}}\right\} \tag{6.53}$$

In each of the double Inequalities 6.47 to 6.53, the term on the left side is the lower (two-tailed) confidence limit and the term on the right side is the upper confidence limit. An explanation of the meaning and appropriate values of the term $t_{\alpha/2,v}$ has already been given in Section 3.1.

3.3 The three-parameter lognormal distribution

(a) Properties of the population distribution

The distribution of a variate X may be such that only when the variate is displaced by a constant τ will a logarithmic transformation, producing the variate $Y=\ln(X-\tau)$, have a normal distribution with mean μ and variance σ^2. Under these circumstances the frequency distribution of $X-\tau$ rather than of X itself, is lognormal. The constant is then taken to be a third parameter which has to be estimated (unless it is determined on *a priori* grounds). It defines the lower bound to the range of values of the variate X and can have either a positive or a negative value which in effect shifts the frequency curve either to the right or to the left relative to the origin. The probability density function for the distribution of $X-\tau$ is thus written as

$$f(x)=\frac{1}{\sigma(x-\tau)\sqrt{2\pi}}\exp\left\{-\frac{1}{2\sigma^2}(\ln(x-\tau)-\mu)^2\right\} \tag{6.54}$$

for $\tau<x<\infty$

where μ and σ^2 are the mean and variance of the distribution of Y. Note that $f(x)=0$ when $x<\tau$.

The mean and median of the distribution of X are respectively at

$$x=\tau+\theta \tag{6.55}$$

and

$$x=\tau+\exp\{\mu\} \tag{6.56}$$

where θ is as defined by Equation 6.26.

Following Equations 6.32 and 6.33 the values of x corresponding to quantiles of the distribution are given by,

$$\text{for 50\% quantile and below,}\quad Q'_{100\alpha}=\tau+\exp\{\mu-Z_\alpha\sigma\} \tag{6.57}$$

$$\text{for 50\% quantile and above,}\quad Q'_{100(1-\alpha)}=\tau+\exp\{\mu+Z_\alpha\sigma\} \tag{6.58}$$

When τ has a negative value, the displacement of the frequency curve may be such that a portion is to the left of the origin. Thus when applied to air pollutant concentrations, which cannot be negative, that part of the curve must be censored and the area under it must be represented by a discrete probability at the origin. An example of a censored three-parameter lognormal frequency distribution is given in the next sub-section.

(*b*) *Estimation of the parameters* τ, μ *and* σ^2

The two simplest methods of estimating the parameters τ, μ and σ^2, from large samples, are described here. The first is the method of quantiles using in particular the 5%, 50% and 95% quantiles. The estimates denoted by c, y_m *and* S_y^2, are determined from:

sample standard deviation,

$$S_g = \frac{1}{1.645}\left\{\ln(q_{95} - q_{50}) - \ln(q_{50} - q_5)\right\} \qquad (6.59)$$

sample mean,

$$y_m = \ln(q_{50} - q_5) - \ln(1 - \exp\{-1.645\, S_y\}) \qquad (6.60)$$

sample displacement parameter,

$$c = q_{50} - \exp\{y_m\} \qquad (6.61)$$

where q_5, q_{50} and q_{95} are the 5%, 50% (median) and 95% sample quantiles.

The second method of estimation is a graphical trial and error process in which different values of the estimator c of the parameter τ are added to the sample observations of X until one is found which produces a straight line on a log-probability plot of cumulative relative frequency data. Estimates y_m and S_y^2 of the parameters μ and σ^2 are then obtained from the straight line using the 16%, 50% and 84% quantiles in the following expressions:

$$y_m = \ln q_{50} \qquad (6.62)$$

$$S_y = \ln\left\{\frac{1}{2}\left(\frac{q_{50}}{q_{16}} + \frac{q_{84}}{q_{50}}\right)\right\} \qquad (6.63)$$

These and other methods of estimation have been described elsewhere.[7] The problems of establishing the efficiency and precision of estimators of τ, μ and σ^2 are a good deal more complicated than those encountered with the two parameters of the simpler lognormal distribution (Section 3.2) and are not considered here.

Example:

As an example of the estimation of τ, μ and σ^2, consider the artificial frequency data on 8-h time averaged pollutant concentrations that are given in Table 6.3. A log-probability plot of the cumulative frequencies (curve A in Fig. 6.9) produced a curve rather than a straight line, indicating that the data did not follow a two-parameter lognormal distribution. Estimates of τ, μ and σ^2 were made by the method of quantiles and by the graphical trial and error procedure in order to try the fit of a three-parameter lognormal distribution to these data. The estimates are given in Table 6.4. The fit of the three-parameter lognormal frequency curve based on the

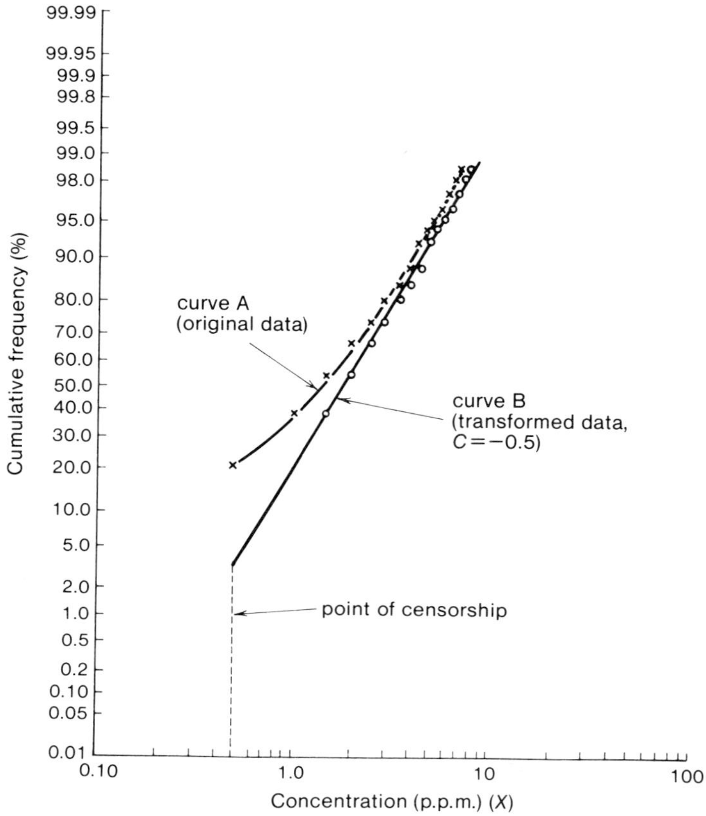

Fig. 6.9 Log-probability plot of pollutant concentrations showing the effect of applying the transformation $X' = X + 0.5$.

estimates determined by the method of quantiles is shown in Fig. 6.10. The point of censorship is at $X - \tau = 0.49$, which corresponds to the 3.6% quantile of the distribution.

4 Goodness of fit tests

The goodness of fit of a theoretical probability density function to an observed frequency distribution of pollutant concentrations should be tested to determine whether it is in fact the most suitable function to adopt as a model for prediction and control in a particular environment, because as shown in the next subsection, there is no reason to assume that any single theoretical distribution is an inherent property of airborne pollutant concentrations, even though one particular form, the lognormal distribution, may be found to have wide application in practice. Two goodness of fit tests that are of general interest for this purpose are the well known chi-square test[13] and the Kolmogorov–Smirnov test.[14,15] Both require large samples.

The chi-square test involves classification of the observed data into K class

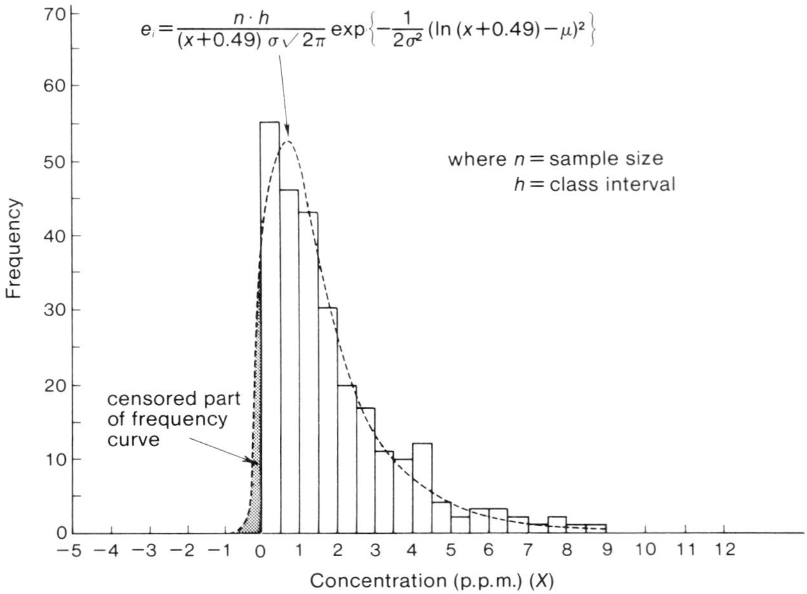

Fig. 6.10 Comparison of observed frequency histogram of pollutant concentrations with expected frequencies, e_i, given by the censored three-parameter lognormal probability model for $\mu=0.623$, $\sigma=0.696$ and $\tau=-0.49$.

intervals and constructing a table of observed frequencies f_i and expected frequencies e_i, the latter being calculated from the theoretical or formal probability density function that is under consideration. For each class interval, a measure of the discrepancy between the observed and expected frequencies is given by $1/e_i(f_i-e_i)^2$ and the measure of the total discrepancy for the observed sample as a whole is the sum of the measures for each interval, which under the hypothesis that the theoretical distribution does fit the observed data, would have a chi-square distribution with either $\nu=K-1$ or $\nu=K-1-\mathrm{d}$ degrees of freedom. The former number of degrees of freedom is used when population parameters of the distribution are known, and the latter is used when d parameters have to be estimated from the observed sample data. Thus

$$\chi^2 = \sum_{i=1}^{K} \frac{1}{e_i}(f_i-e_i)^2 = \sum_{i=1}^{K} \frac{f_i^2}{e_i} - n \tag{6.64}$$

There are two methods of constructing the frequency table for the chi-square test. In both cases the aim should be to avoid having intervals with frequencies of less than 5. One method is to use equal class intervals and to calculate the expected frequencies from

$$e_i = nP_i = n\int_{x_a}^{x_b} f(x)\,\mathrm{d}x \tag{6.65}$$

where n is sample size, P_i is expected probability for ith interval, x_a, x_b are lower and

Table 6.3 Cumulative frequencies of 8-h time averaged pollutant concentrations experienced by a group of employees over a period of one year (artificial data)

Concentration interval, p.p.m.	*Mid interval* (x_i)	*Frequency*	*Cumulative relative frequency* (%)
0–0.5	0.25	55	20.91
>0.5–1.0	0.75	46	38.40
>1.0–1.5	1.25	43	54.75
>1.5–2.0	1.75	30	66.16
>2.0–2.5	2.25	20	73.81
>2.5–3.0	2.75	17	80.27
>3.0–3.5	3.25	11	84.45
>3.5–4.0	3.75	10	88.26
>4.0–4.5	4.25	12	92.82
>4.5–5.0	4.75	4	94.34
>5.0–5.5	5.25	2	95.10
>5.5–6.0	5.75	3	96.24
>6.0–6.5	6.25	3	97.38
>6.5–7.0	6.75	2	98.14
>7.0–7.5	7.25	1	98.52
>7.5–8.0	7.75	2	99.28
>8.0–8.5	8.25	1	99.66
>8.5–9.0	8.75	1	100
		$N = 263$	

upper limits of ith interval and $f(x)$ is the probability density function of interest.

It may not be possible to avoid having frequencies of less than 5 for class intervals in the tails of the distribution. Under these circumstances certain class intervals have to be pooled for the purposes of the chi-square test.

Example:

The equal interval frequency data in columns 1 to 3 of Table 6.5 refer to 101 consecutive 3-min average styrene-in-air concentrations obtained from curve B in Fig. 6.4. The mean and standard deviation of logarithmic transformations of the concentration values are $\bar{y} = 3.4171$ and $S_y = 0.6074$. It is hypothesised that the frequency distribution is lognormal and the χ^2 test is used to test the goodness of fit of a two-parameter lognormal distribution defined by the observed mean and variance.

Table 6.4 Estimated values of parameters of the distribution of $X - \tau$

Parameter	*Estimator*	*Method of quantiles (Equations 6.59 and 6.61)*	*Graphical method (Curve B, Fig. 6.9)*
τ	c	−0.49	−0.50
μ	y_m	0.6259	0.5933
σ	S_y	0.696	0.691

Table 6.5 Calculation of expected frequencies for styrene-in-air data assuming a two-parameter lognormal distribution

1	2	3		4	5	6	7	8	9	10		11	12
Styrene concen-tration interval $x_a - x_b$	*Mid interval value* x_i	*Observed frequency* f_i		*Log-interval* $y_a - y_b$	*Standard normal variate* Z_a $\left[\frac{y_a - \bar{y}}{s}\right]$	*Standard normal variate* Z_b $\left[\frac{y_b - \bar{y}}{s}\right]$	$\Phi(Z_a)$ *Area under standard normal curve (from tables)* 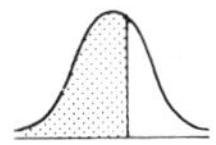 Z_a	$\Phi(Z_b)$ *Area under standard normal curve (from tables)* 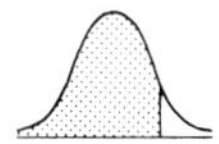 Z_b	$\Delta\Phi$ 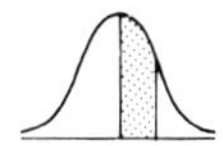 $Z_a Z_b$	*Expected frequency* e_i $[n \cdot \Delta\Phi]$		f_i^2/e_i	*Height of expected frequency curve for mid interval values,* x_i [n.h. $f(x)$]
0–10	5	3	24	<2.30	—	−1.8392	—	0.03294	0.03294	3.33	24.9	23.17	1.58
10–20	15	21		2.30–3.00	−1.8392	−0.6878	0.03294	0.24614	0.2132	21.53			22.38
20–30	25	27		3.00–3.40	−0.6867	−0.02808	0.24614	0.48880	0.24266	24.51		29.74	25.16
30–40	35	15		3.40–3.69	−0.02808	0.4494	0.48880	0.67343	0.18463	18.65		12.06	18.47
40–50	45	14		3.69–3.91	0.4494	0.81163	0.67343	0.79150	0.11807	11.93		16.43	12.00
50–60	55	8	13	3.91–4.09	0.81163	1.10800	0.79150	0.86607	0.07457	7.53	12.5	13.56	7.52
60–70	65	5		4.09–4.25	1.10800	1.37144	0.86607	0.91500	0.04881	4.93			4.69
70–80	75	3		4.25–4.38	1.37144	1.58549	0.91500	0.94357	0.02869	2.90			2.95
80–90	85	3		4.38–4.50	1.58549	1.78306	0.94357	0.96271	0.019138	1.93			1.88
90–100	95	1		4.50–4.61	1.78306	1.96418	0.96271	0.97525	0.01254	1.27			1.21
100–110	105	0	8	4.61–4.70	1.96418	2.11236	0.97525	0.98267	0.00743	0.75	8.2	7.84	0.79
110–120	115	0		4.70–4.79	2.11236	2.26055	0.98267	0.98811	0.00544	0.55			0.53
120–130	125	0		4.79–4.87	2.26055	2.39227	0.98811	0.99163	0.00352	0.36			0.36
130–140	135	0		4.87–4.94	2.39227	2.50752	0.99163	0.99392	0.00229	0.23			0.24
140–150	145	1		4.94–5.01	2.50752	2.62278	0.99392	0.99564	0.00172	0.17			0.17
> 150		0		>5.01	2.62278	—	0.99564	—	—	—		—	—

Steps in the calculation of the expected frequencies for each logarithmic class interval are shown in columns 5 to 9 of Table 6.5. Columns 6 and 8 are effectively redundant and are shown only for the sake of clarity. Values of the distribution function $\Phi(Z)$ which is defined by the integral of Equation 6.7 between the limits of $-\infty$ and Z, may be obtained from published tables or, more conveniently, from an electronic calculator or computer which is suitably programmed. The height of the expected frequency curve (column 12) at each of the mid-interval values, x_i, is determined by evaluating Equation 6.25 then multiplying by the total number of observations, n, times the width of the class intervals, $h = x_b - x_a$. It can be seen that some pooling of frequencies is necessary to avoid expected frequencies of 5 or less.

The sum of values of f_i^2/e_i is 102.8, so that from Equation 6.64,

$$\chi^2 = 102.8 - 101 = 1.8$$

With pooling there are $K = 6$ class intervals, so that the number of degrees of freedom are given by

$$\nu = K - 1 - \mathrm{d} = 3$$

For three degrees of freedom, the 5% point on the χ^2 distribution is at $\chi^2 = 7.815$. This is higher than the observed value, thus it is inferred that the latter is not significant (at the 5% level) and that the initial hypothesis of lognormality of the data in this example is correct.

The second method of constructing the frequency table for the purposes of a chi-square test, which might be simpler, is to select unequal class intervals which correspond to equal probabilities and hence equal frequencies. Thus if K is the number of class intervals, the probability of each interval is

$$P_i = 1/K \quad \text{for} \quad i = 1, 2, \ldots K \tag{6.66}$$

and the expected frequency for each class interval is

$$e_i = P_i n = n/K \tag{6.67}$$

Now P_i represents the area under a portion of the curve of the theoretical probability density function $f(x)$, so that the class interval limit K_i can be determined by successive integrations of the function between limits which give the same probabilities, using either estimates or known values of the population parameters. In the case of the normal distribution, for example, the probability density function for the standardised normal distribution, Equation 6.7, is integrated between the limits $-\infty$ and Z_i (yielding the distribution function $\Phi(Z)$) and is then equated to iP_i to produce limit values Z_i of the standard normal variate. The procedure is simple in this case since tables of areas under the standardised normal curve can be used. The limits, K_i, are quantiles and can thus be determined by substituting absolute values of Z_i and values of the estimates of the parameters μ and σ into Equations 6.9 or 6.10.

Example:

If $n = 1000$ and $K = 10$ so that each $P_i = 0.1$, and if $\bar{x} = 10$ and $S = 2.0$, then for the purposes of a chi-square test on the goodness of fit of the normal distribution, the required class interval limits would be those given in Table 6.6.

Where the probability density function of interest is that of a two-parameter lognormal distribution or that of a three-parameter lognormal distribution, the standardised normal distribution again can be used to determine equal probability class intervals limits Z_i, bearing in mind that the respective transformations $Y = \ln X$ and $Y = \ln(X-\tau)$ for the two distributions, are normally distributed. The values of Z_i for 10 class intervals would be as given in column 3 of Table 6.6. Using a similar procedure to that used in the above example, the class interval limits K_i in terms of the variate X would then be given by Equations 6.32 and 6.33 in the case of the two-parameter lognormal distribution and by Equations 6.57 and 6.58 in the case of the three-parameter lognormal distribution.

The second goodness of fit test that was mentioned at the beginning of this section, the Kolmogorov–Smirnov test,[14,15] is based on the maximum absolute difference between the observed cumulative frequency distribution, $S_n(x)$, and the theoretical cumulative frequency distribution, $F_n(x)$. The test requires, initially, the construction of a frequency table of observed and expected frequencies and of cumulative frequencies, for equal class intervals. The test statistic is thus,

$$\frac{d}{n} = \max \left| \frac{F_n(x) - S_n(x)}{n} \right| \tag{6.68}$$

The value of d/n is compared with the tabulated[14] critical value, $d_{\alpha,n}$, corresponding to the sample size n and the chosen level of significance, α. If $d/n \geq d_{\alpha,n}$ then it is inferred that the observed data do not belong to the hypothesized distribution.

A graphical method can be used for testing the goodness of fit of a normal

Table 6.6 Example of calculated class interval limit values for chi-square test on the goodness of fit of a normal probability density function

Number of class interval (i)	*Cumulative probability of class intervals where* $p_i = 0.1$ (iP_i)	*Interval limit of standard normal variate* (Z_i)	*Interval limit*[1] *of variate X* (K_i)
1	0.1	−1.282	7.436
2	0.2	−0.842	8.316
3	0.3	−0.524	8.952
4	0.4	−0.253	9.494
5	0.5	0	10
6	0.6	0.253	10.506
7	0.7	0.524	11.048
8	0.8	0.842	11.684
9	0.9	1.282	12.564
10	1.0	∞	∞

[1] (*a*) for values of $iP_i \leq 0.5$ use Equation 6.9;
(*b*) for values of $iP_i \geq 0.5$ use Equation 6.10.

distribution. This is based on the fact that cumulative frequencies of a normal distribution produce a straight line when plotted on probability graph paper. A similar procedure can be used for testing lognormality. In this case, cumulative frequencies of a lognormal distribution produce a straight line when plotted on log-probability graph paper. The graphical approach is simple and convenient to use, but it is not a rigorous statistical test. It can be subjective to a certain extent and is open to misinterpretation which might lead to an incorrect decision about the true form of a distribution.

5 Application of theoretical functions to observed frequency distributions

5.1 Within shift variation

It has been assumed[16–18] on the basis of fairly limited published data[19,20] that the two-parameter lognormal distribution is a good general descriptor of the variability of pollutant concentrations experienced by an individual employee over the period of a shift of, say eight hours duration, and that this distribution can thus be used as a general model for making decisions about compliance with exposure limits, such as 8-h time weighted average limits, from short-time air sampling data. However, the general applicability of the two-parameter lognormal model to within-shift variation at a fixed point or in the breathing zone of an employee (over the whole shift) should be viewed with scepticism – it should certainly not be taken to be an inherent property of such data.

Central limit theory says that a random variate X is lognormally distributed if it is the product of a number of mutually independent random processes. The differences observed in the results of a sequence of short-term personal air samples or in moment-to-moment values of pollutant concentration indicated by a continuous measuring device located in the breathing zone, are not likely to arise from elemental processes which are strictly independent or indeed random, so that a lognormal distribution of the observed concentrations is not an inevitable consequence of such processes. Even so, it may turn out to be a good empirical fit to the data under some circumstances.

In fact, the degree of positive skew and other aspects of the shape of the frequency distribution of within shift pollutant concentrations experienced by a worker can vary considerably depending upon the standard of pollutant control, the nature of the work being done and the general pattern of exposure over the assessment period. For example, if the pattern of work is such that a person is exposed to relatively high levels of a pollutant for short periods, with fairly long intervening periods of exposure to low levels, or if the nature of the work varies systematically over an 8-h shift, then the observed frequency distribution of pollutant concentration for the 8-h assessment period could be J-shaped. This is illustrated in Fig. 6.11(*a*) and (*b*) which are the histogram and cumulative frequency curve for the distribution of 160 consecutive 3-min average styrene concentrations derived from the continuous exposure curve presented in Fig. 6.3. From the shape of the log-probability curve of

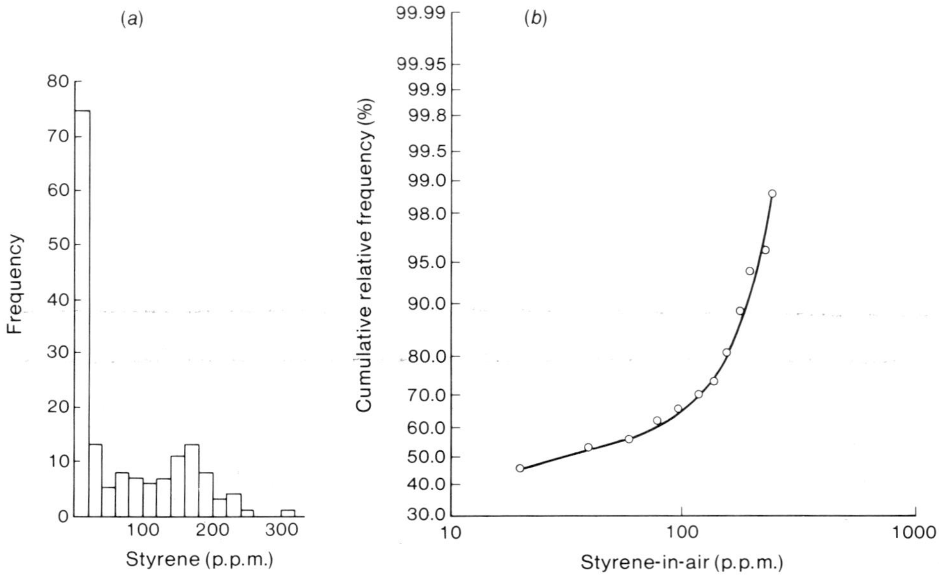

Fig. 6.11 Distribution of 3-min average styrene concentrations over an 8-h shift. (*a*) Histogram: arithmetic mean, $\overline{X} = 68.8$ p.p.m.; geometric mean, $M_g = 8.6$ p.p.m.; geometric standard deviation, $S_g = 29$ p.p.m.; $N = 160$. (*b*) Cumulative frequency curve.

cumulative frequencies it is clear that the distribution is far from lognormal.

In the case of a person in a job which involves the same kind of activity in the same working zone for the duration of a given assessment period (which may be the whole or part of an 8-h shift) and where there is no time trend in pollution levels due to systematic changes in one or more of the elemental processes involved, it may be found that the lognormal distribution is a good descriptor of the variability of pollutant concentrations over that assessment period. Even where there is a discernible time trend in the pollutant levels, the lognormal distribution may be found to be an acceptable fit occasionally, but it would have doubtful value and reliability as a model for decision making under such circumstances.

For examples of frequency distributions arising from the kind of situation that has just been described, and of how the shape of frequency distributions can be affected by ventilation control conditions, we can refer back to the two continuous styrene-in-air recordings compared in Fig. 6.4(*a*) and (*b*). The frequency distributions of styrene concentrations taken at 15-s intervals over the 25-min assessment period are compared in Fig. 6.12.

For the local exhaust ventilation condition, case B, a two-parameter lognormal distribution was found to be a good fit to the observed frequency distribution of styrene concentrations ($\chi^2_{\text{obs}} = 1.8$ with $\nu = 3$; $0.5 < p < 0.7$). For the poor ventilation condition, case A, the lognormal distribution was a poorer fit due to the upward time trend in the data, but was not rejected by the chi-square goodness of fit test at the 5% significance level ($\chi^2 = 14.5$ with $\nu = 8$; $0.05 < p < 0.10$). The poor

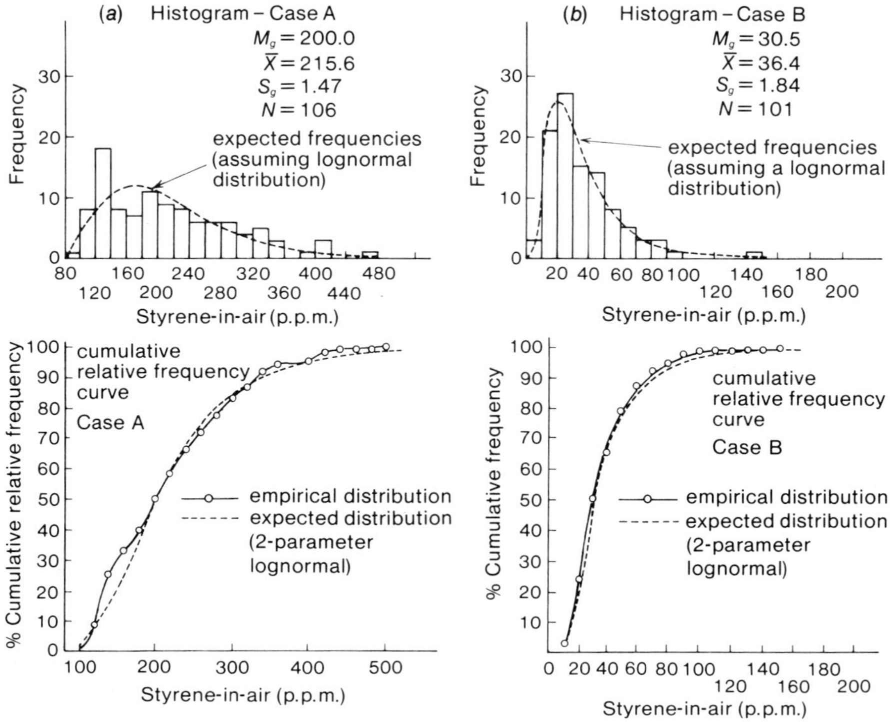

Fig. 6.12 Distribution of styrene-in-air concentrations in the breathing zone of a person hand-laminating the third ply of a 4.5 m long GRP sailing boat hull under: (*a*) poor general ventilation conditions; (*b*) local exhaust ventilation conditions. The data were taken from the continuous recordings shown in Fig. 6.4.

ventilation condition led to a higher mean M_a and variance S_a^2 (estimates of θ and β^2 determined from Equations 6.45 and 6.27 respectively) than the local exhaust ventilation condition.

5.2 Between-shift variation

There appears to be rather more evidence in the literature to support claims that the variation of time weighted average concentrations measured over long assessment periods on a shift (e.g. 8-h time weighted average), daily or weekly basis can be adequately described for practical purposes by the two-parameter lognormal distribution.[1,21–24] This may often be found to be so in the case of time weighted average concentrations measured for groups of workers by personal sampling and of those measured for general workroom environments by multipoint static sampling, where the same general pattern of work activity and plant or process operation is repeated day after day.

Studies of atmospheric pollution near urban industrial areas by Larsen[25–27] have indicated that here too there is lognormality of distributions of pollutant concentrations. Furthermore, Singpurwalla[28] has provided a theoretical statistical interpretation of Larsen's findings. Such evidence together with that cited above for working environments has been taken by some investigators to confirm the general applicability of the two-parameter lognormal model. However, a number of other studies have indicated that it would be unwise to assume automatically that the two-parameter lognormal probability model applied to a particular situation.[29–32] The caution that was advocated in the previous subsection should also be exercised when dealing with distributions of full shift or longer term average concentrations of a pollutant.

An analysis of published data by Ott and Mage[33,34] has shown that the three-parameter lognormal distribution is usually a better fit to pollution data than the two-parameter lognormal distribution. Ott and Mage suggested that this provides a more flexible model and may have a much wider application than the two-parameter model. The two-parameter lognormal distribution can be regarded, in fact, as a special case of the three-parameter distribution, which is obtained when parameter τ is equal to zero.

Two examples of observed distributions of shift average (8-h) concentrations of pollutants in working environments are shown in Figs. 6.13 and 6.14. The data in Fig. 6.13 are average concentrations of a pollutant vapour in a chemical process area (indoors), measured over 1176 consecutive 8-h shifts, three shifts per day, by a multipoint, continuous-sequential monitoring system. The two-parameter lognormal distribution was not a good fit at the 5% significance level by a chi-square test ($\chi^2_{obs} = 74.8$, $\nu = 14$), but a censored three-parameter lognormal distribution was a good fit ($\chi^2_{obs} = 11.9$, $\nu = 11$; $0.3 < p < 0.5$). The comparison of observed and predicted values of the median and of the 95% quantile in Table 6.7 shows that for this example the percentage error of the expected value of the median given by either model was not more than 4% (values given in parentheses in the Table) and that the error of the 95% quantile given by the three-parameter model was also low. However, the percentage error between the observed 95% quantile and the expected value given by the two-parameter model was almost 15%. The practical significance of such a discrepancy may depend, among other things, upon the accuracy of the method used for measuring pollutant concentrations.

Table 6.7 Comparison of observed and expected values of the median and 95% quantile for the data in Fig. 6.13

Statistic	*Observed value*	*Expected value given by two-parameter lognormal model*	*Expected value given by three-parameter lognormal model*
Median	2.02	1.95 (3.5%)	2.10 (4.0%)
95% quantile	5.2	5.96 (14.6%)	5.12 (1.5%)

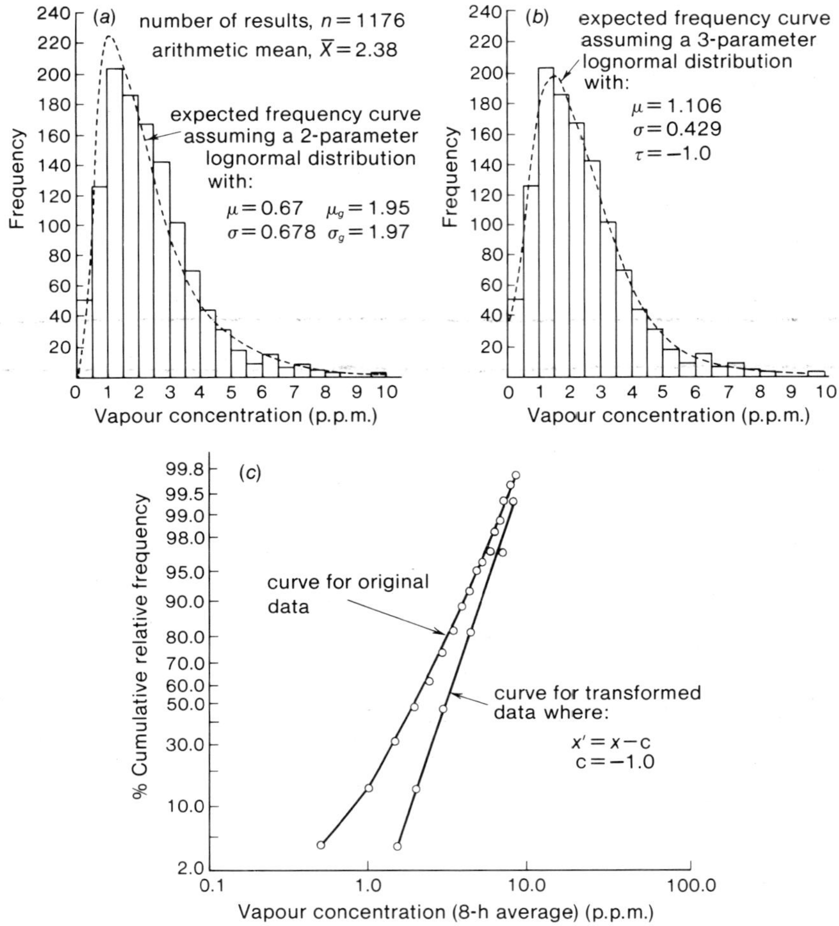

Fig. 6.13 Distribution of full shift (8-h) average concentrations of a pollutant vapour in a chemical process area (1176 consecutive shifts).

The data in Fig. 6.14 are 8-h average exposure levels of dust and fume experienced by workers involved in an industrial process. They were obtained from thirty-nine full shift personal sampling trials made at varying intervals over a 24-month assessment period. In this case a two-parameter lognormal distribution was a good fit to the data ($\chi^2_{obs} = 13.6$, $\nu = 10$; $0.1 < p < 0.2$).

5.3 Distribution of sample means

It was stated earlier that under the central limit theorem the sampling distribution of the means of random samples is normal, or approximately so. This is demonstrated here by an example which makes use of the skew parent distribution of 1920 values of 'instantaneous' styrene-in-air concentrations taken at 15-s intervals from the 8-h

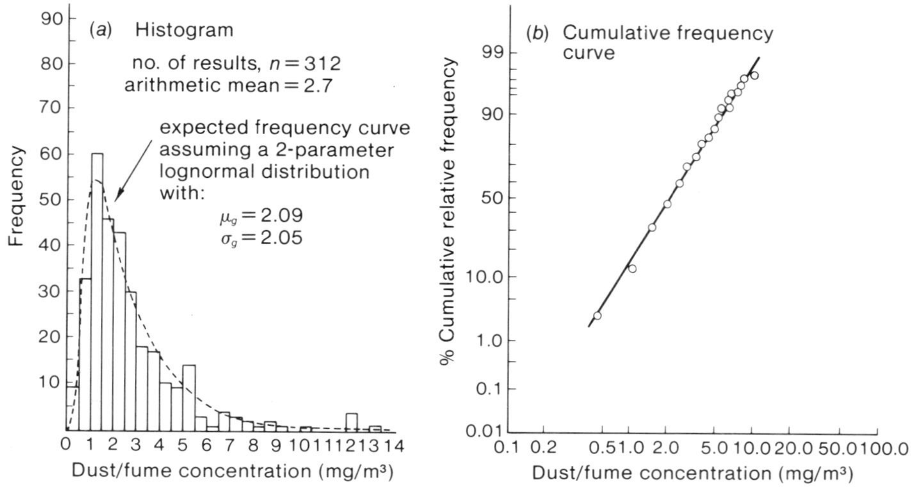

Fig. 6.14 Frequency distribution of full shift (8-h) personal exposures of a group of employees to dust and fume.

continuous recording shown in Fig. 6.3. The frequency distribution of these values is shown in Fig. 6.15(*a*).

In a sampling experiment using a computer, 200 random samples of size $n = 100$ were drawn (with replacement) from the parent population. Each element of each sample was a 3-min average concentration determined by calculating the arithmetic mean of twelve consecutive instantaneous styrene concentration values from a randomly selected starting point. The mean of the 100, 3-min averages for each sample was calculated to provide in each case an estimate of the true 8-h time

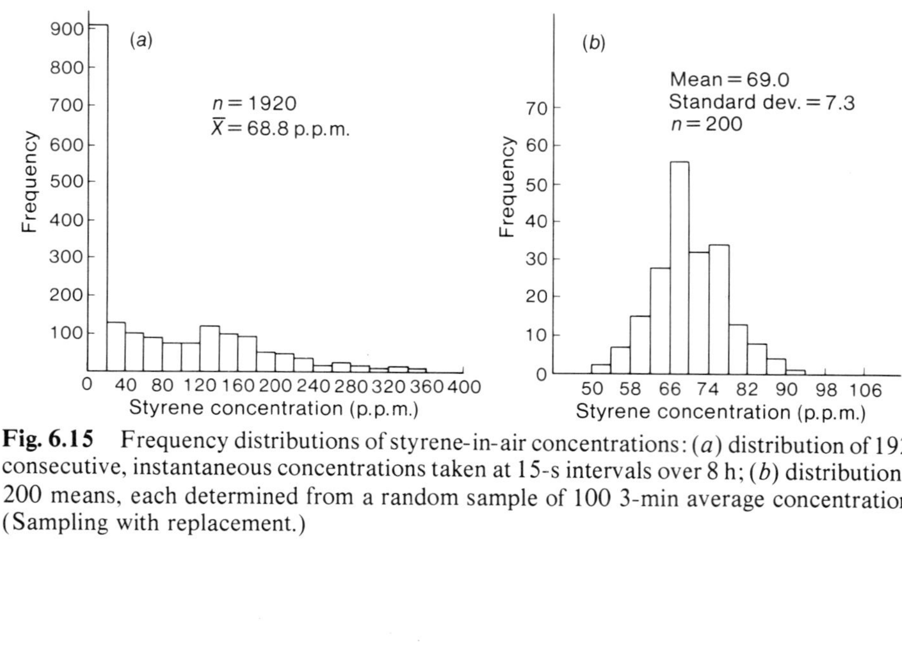

Fig. 6.15 Frequency distributions of styrene-in-air concentrations: (*a*) distribution of 1920 consecutive, instantaneous concentrations taken at 15-s intervals over 8 h; (*b*) distribution of 200 means, each determined from a random sample of 100 3-min average concentrations. (Sampling with replacement.)

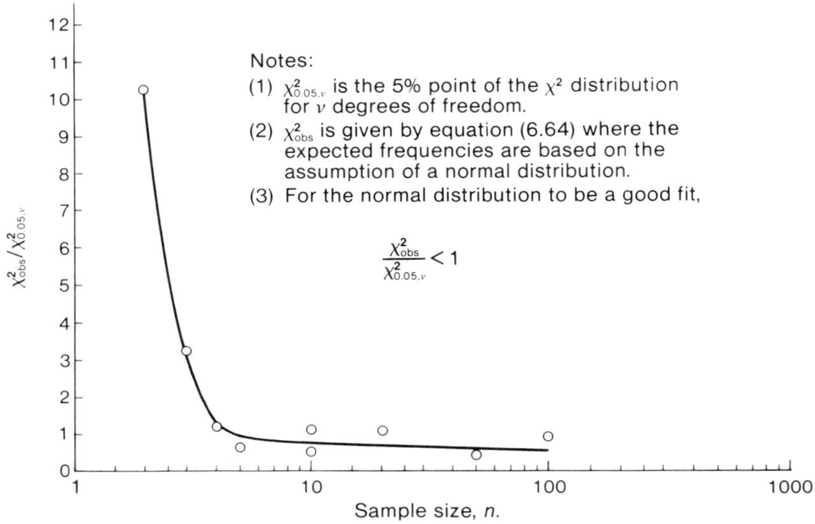

Fig. 6.16 Results of χ^2 test on the goodness of fit of a normal distribution to the frequency distribution of the means of 200 samples of size *n*.

weighted average. The frequency distribution of the 200 sample means so calculated, is shown in Fig. 6.15(*b*).

The procedure was repeated for random samples of sizes $n = 2, 3, 4, 5, 10, 20$ and 50. The change in shape of the sampling distributions of means from skew to normal, as the sample size, *n*, was increased is illustrated in Fig. 6.16. The graph shown is a plot of *n* versus the ratio of the observed value of chi-square calculated from expected frequencies on the assumption of a normal distribution, to the 5% point of the chi-square distribution (i.e. χ^2 goodness of fit test). The graph indicates that for this example, the means of samples of size $n \geq 5$ are normally distributed, or approximately so.

The relationship between sample size, *n*, and standard deviation, $S_{\bar{x}}$, of the sampling distribution is shown in Fig. 6.17. The relationship between *n* and the geometric standard deviation, S_g, is shown in Fig. 6.18. Both curves show that the greatest rate of change of either measure of dispersion with respect to sample size, occurred at $n < 30$. This suggests that for the optimum precision of estimation of the true mean styrene concentration for the 8-h assessment period, the number of 3-min styrene-in-air measurements made at random over that period would need to be equal to or greater than 30. This may be a good rule to apply generally to the problem of estimating full shift average exposures to pollutants from very short time sampling devices such as gas detector tubes.

6 Effects of averaging time

As the averaging time of consecutive air samples covering the whole of an assessment period is increased, so it is found that there is smoothing of the variation

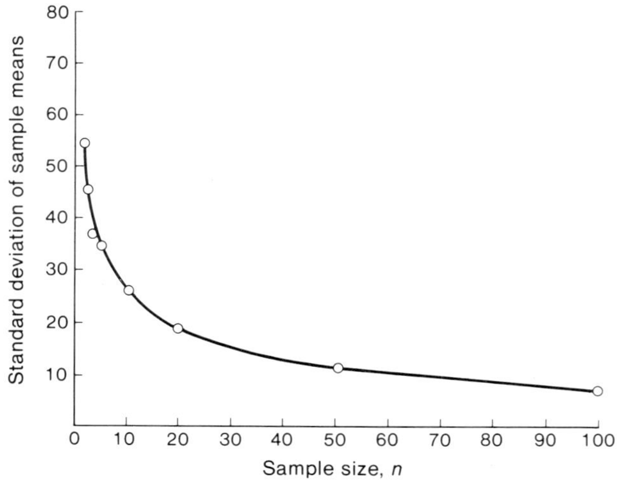

Fig. 6.17 The effect of sample size on the standard deviation, $S_{\bar{x}}$, of the distribution of 200 sample means.

in the measured concentrations of a pollutant. This means, of course, that the information value of the longer term and hence fewer samples is progressively reduced in the sense that they may not detect transient patterns of change or more persistent time trends in the exposure of a worker or in emissions from a process during the assessment period, which may be of considerable interest to the occupational hygienist or to the engineer concerned with environmental control. The averaging time is thus an important factor to be considered when deciding upon the overall strategy to be used for a particular occupational hygiene survey or for an air monitoring programme.

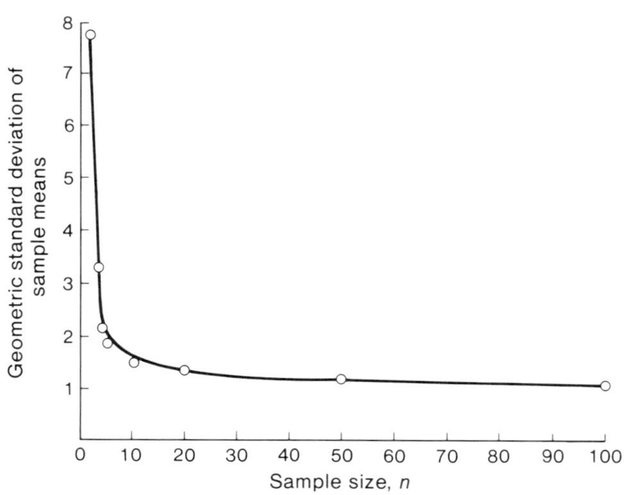

Fig. 6.18 The effect of sample size on the geometric standard deviation of the distribution of 200 sample means.

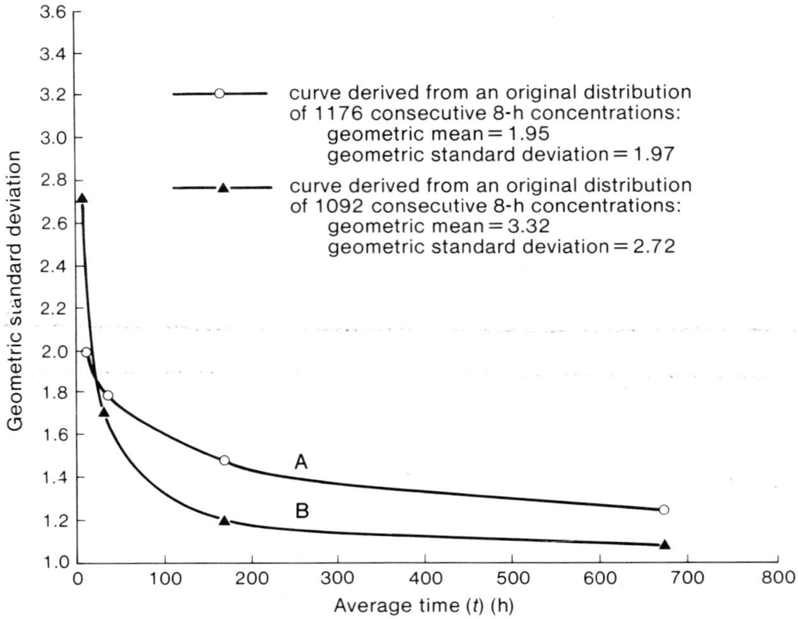

Fig. 6.19 The effect of averaging time on the geometric standard deviation of average concentrations of pollutants determined by consecutive air sampling over long assessment periods.

Fig. 6.19 shows two examples of the effect of averaging time on the geometric standard deviation S_g of the distribution of n consecutive time averaged concentrations. The two curves were derived by calculating, from 8-h shift data in either case, the time averaged pollutant-in-air values and the geometric standard deviations for n_1 consecutive 24-h (daily) periods, n_2 consecutive 168-h (weekly) periods and n_3 consecutive 672-h (4-weekly) periods.

Saltzman[35] has suggested the following empirical relationship between averaging time t and geometric standard deviation S_g, which was derived from an analysis of urban pollution data:

$$\ln S_g = \ln K - m \ln t \tag{6.69}$$

where S_g is the geometric standard deviation, t is the averaging time in hours, K is the value of S_g when $t = 1$, m is the slope.

There is, however, some doubt about the generality of this relationship. With regard to the two curves shown in Fig. 6.19, only one of them, curve A, would be linearised by plotting the data on logarithmic scales.

Where continuous air monitoring at a continuously operating chemical process plant is done by static sampling and where, for the pollutant concerned, a target value for the long term (one year or more) average concentration, θ, has been established, there may be some advantage to be gained by analysing the monitoring data in the way that has been described, to determine the relation between t and S_g. The kind of information given by curves such as those in Fig. 6.19 might be of value, for example, in deciding upon the optimum averaging time to use in the future for the purpose of detecting any deterioration in control of the pollutant which may influence the long term average concentration.

Suppose that for averaging times, t, of between 1 h and 168 h, concentrations of the pollutant in the process area are found to be lognormally distributed and suppose that a particular value of t within this range is chosen for the purposes of monitoring the control of emissions. The corresponding value of the geometric standard deviation, S_g, can be read off a curve such as A or B in Fig. 6.19 and this can be entered together with the target average concentration, θ, into Equation 6.37 to estimate, say, the 95% quantile value of concentration. The latter can then be used as a simple form of control limit for the chemical process area in question.

When an estimate of the time averaged concentration for an assessment period T hours is made by obtaining air samples at random or systematic intervals so that there are unsampled gaps within the assessment period, the precision of estimation improves, in fact, as the size of the ratio T/t is decreased towards one.

7 Measuring error

A number of sources of variation in observed concentrations of airborne pollutants were described earlier in this chapter. It should be recognised that in addition to these, measuring error can contribute to significant differences between, for example, repeated measurements of concentrations at the same location and between measurements made by different techniques and equipment. Measuring error may be the sum of a number of independent components arising from the air sampling and the analytical techniques that are used. It may be due to systematic error (or bias as it is often called), random error, or a combination of both types.

Systematic error in measurements of pollutant concentrations, which defines the accuracy of the methods used, is a problem that is easily overlooked. However, it can sometimes be more serious than the problem of random errors, which tend to cancel out as the number of measurements made in a particular situation, is increased. It is quite possible, for example, for two laboratories, A and B, using different sampling and analytical methods (or even nominally the same methods) to achieve individually the same degree of precision in repeated measurements of a particular gas-in-air concentration, but for the mean value determined by laboratory A to differ substantially from that determined by laboratory B. It is also possible for the mean values determined by both laboratories to differ from the true concentration.

It is important, therefore, to ensure either that no systematic errors exist in the sampling and analytical techniques that are used for a particular pollutant, or at least to identify them and estimate their magnitude so that correction factors can be applied to allow for them. Differences between types of sampling device (including the pumps), between different analytical methods, between analysts and between laboratories, and comparisons with recognised standards, can be investigated by statistically designed experiments which permit the use of simple comparisons, analysis of variance and regression analysis methods to separate and estimate the statistically significant systematic errors.[4,36–38]

Random error would affect the precision of repeated measurements of a non-varying gas concentration, that is the closeness with which the measurements agreed with one another. To state the problem of random error rather simply, we can first write that for an individual measurement, x_i, of that gas concentration,

$$x_i = x_0 + e_i \tag{6.70}$$

where x_0 is the true concentration and where e_i is the measuring error which is assumed to be a normally distributed random variable with a population mean of zero and variance σ_e^2. For n repeated measurements, we can then write that,

$$\bar{x} = x_0 + \frac{1}{n}\sum_{i=1}^{n} e_i \tag{6.71}$$

The precision of the estimate, $\bar{x}$, of the true concentration x_0, in the case of the non-varying concentration, would depend upon the error variance σ_e^2 and upon the number of repeated measurements, n. This can be expressed in statistical terms by reference to the two-tailed $100(1-\alpha)\%$ confidence limits (see Expression 6.15) given by

$$\bar{x} \pm Z_{\alpha/2} \cdot \frac{\sigma_e}{\sqrt{n}} \tag{6.72}$$

In the measurement of pollutant concentrations the random error variate, e, is likely to be the sum of independent random components such as sampling pump flow rate error, losses of the pollutant chemical from the collecting medium, fluctuations in the performance of analytical instrumentation, etc., so that the assumption that it is normally distributed is not unreasonable and indeed would be predicted from central limit theory. The error variance would thus be the sum of r independent components,

$$\sigma_e^2 = \sigma_1^2 + \sigma_2^2 + \ldots + \sigma_r^2 \tag{6.73}$$

Therefore, $$\sigma_e = \sqrt{\sigma_1^2 + \sigma_2^2 + \ldots + \sigma_r^2} \tag{6.74}$$

In terms of coefficients of variation given by

$$\mathrm{CV} = \sigma/\mu \tag{6.75}$$

the combined error can be written as

$$\mathrm{CV}_e = \sqrt{\mathrm{CV}_1^2 + \mathrm{CV}_2^2 + \ldots + \mathrm{CV}_r^2} \tag{6.76}$$

Now, in working environments, the concentrations of pollutants vary over time in the ways which were described earlier. If, therefore, the time averaged concentration, $\bar{X}$, of a pollutant gas at a given site over an assessment period of, say, 8 h were estimated by making n relatively short term measurements at random during that period, then the precision of the estimate, $\bar{x}$, would depend on the actual variability of the instantaneous or short time averaged concentrations as well as on the measuring error and on the sample size, n. If such concentrations were normally distributed with known variance σ^2, then the precision of $\bar{x}$ could be expressed by the $100(1-\alpha)\%$ confidence limits given by

$$\bar{x} \pm Z_{\alpha/2} \frac{\sqrt{\sigma^2+\sigma_e^2}}{\sqrt{n}} \tag{6.77}$$

If the variance σ^2 were not known and thus had to be estimated from the n sample measurements, then $100(1-\alpha)\%$ confidence limits (see Expression 6.21) would be given by,

$$\bar{x} \pm t_{\alpha/2,n-1} \cdot \frac{\sqrt{S^2+\sigma_e^2}}{\sqrt{n}} \tag{6.78}$$

This expression might also be used to obtain an indication of the precision of $\bar{x}$ in the more usual case of a skew distribution of short time averaged concentrations, provided that the sample size n is greater than five.

If the sample size is large or if the variability of concentrations of the pollutant at the sampling site is considerably greater than the random measuring error, which is usually so in working environments, then the effect of σ_e^2 on the confidence limits given by either Expressions 6.77 or 6.78 would be negligible by comparison. On the other hand, if either m consecutive time-averaged concentration measurements of equal duration covering the full assessment period (for example, four consecutive 2-h time averaged measurements over an 8-h period) were made, or if m simultaneous full period measurements were made at the same point, then the measuring error variance, σ_e^2 or its estimate, S_e^2, would be the sole source of uncertainty about the estimate $\bar{x}$ and $100(1-\alpha)\%$ confidence limits would be given by either

$$\bar{x} \pm Z_{\alpha/2} \cdot \frac{\sigma_e}{\sqrt{m}} \quad \text{or} \quad \bar{x} \pm Z_{\alpha/2} \cdot \frac{\mathrm{CV}_e\sqrt{\Sigma x_i^2}}{m}\,\dagger \tag{6.79}$$

8 Air sampling strategies

Attention in this section is confined to air sampling strategies for long-term studies of pollutant concentrations in working environments and to a brief consideration of how these strategies, in terms of the location, number, duration and frequency of air sampling, might be influenced by the statistical aspects of pollutant concentrations. It is recognised, of course, that other factors, too, will have a bearing on the strategy adopted in any particular environment. It is also recognised that occupational hygienists and others often make measurements of pollutant concentrations for reasons other than those considered here, and that they make observations of, and take account of, a variety of factors in addition to air sampling data when assessing risks to health and safety in working environments.

† This alternative applies if the magnitude of measuring error variance varies in proportion to the concentration being measured. The x values here are the results of consecutive measurements.

8.1 Personal sampling

An organisation might undertake a long-term programme of measurement of the exposures of its employees to an air pollutant by personal air sampling, for one or more of the following purposes:

(*a*) *Monitoring* – to keep under surveillance how exposure levels compare with the recognised exposure limit for the pollutant in question. The type of exposure limit considered here is that which is expressed as a time averaged concentration for a workshift. Examples of such limits are the 8-h time weighted average threshold limit values for chemical agents which are published by the American Conference of Governmental Industrial Hygienists[39] and the MAK values of the Federal Republic of Germany.[40]

(*b*) *Monitoring* – to check whether a target long term average exposure concentration of the pollutant is likely to be achieved. The assessment period for averaging might be, for example, one year. A target of this kind might be set by an organisation for purposes which are supplementary to (*a*). It might be connected with the development of environmental improvements or with the use of a statistical approach to control and monitoring.

(*c*) *Data collection* – to obtain data on the exposures of individual employees and of groups of employees over a long period for the purpose of studying correlations of these with data on health. The parameters of interest here may be the long term average exposure concentration of the pollutant and the variability of 8-h shift concentrations. The assessment period might be 1 yr or much longer.

Some of the factors which lie behind decisions to undertake air sampling for these and other purposes are discussed by Miller and Kane in Chapter 4 and by Coker in Chapter 5.

A programme designed to suit one or other of the above objectives would, in general outline, consist of making air sampling trials on a number of carefully defined groups of employees at a frequency of k times per year (k may differ for different groups). On each trial a number of the members of each group would be selected to wear personal samplers in order to estimate their individual full shift time averaged exposure levels. For the purposes of (*b*) and (*c*) the group mean and variance might then be estimated from the individual results obtained on each trial.

In a large organisation, the concept of grouping would be an important one to consider when planning an air monitoring programme for a particular pollutant, since this would help to maximise its effectiveness and efficiency.[41] Grouping would be essential in the case of exposure data collection for epidemiological purposes. In either case, that is for monitoring or for epidemiology, the grouping of employees would require careful preliminary investigation and may, in some cases, need to be a developmental process. It should be done on the basis of similarity of the nature of tasks performed and similarity of certain environmental factors such as the type of process and the method of pollution control for that process. The presence of more than one air pollutant in certain parts of the working environment is another important grouping factor, especially for the purposes of epidemiology.

From the statistical point of view the advantage of grouping would be that the variability of full shift exposure levels would be smaller for well defined groups than that for the exposed workforce as a whole (assuming that this embraced various kinds of work activity and process). As such, the parameters of frequency distributions of shift average exposures would be estimated with more precision and there would be a greater likelihood of being able to describe the variability by means of a formal probability model such as the two-parameter lognormal model. It would also be reasonable in principle to select at random a number of members of a group as a representative sample for the purposes of each trial of the air monitoring programme.

There are three main strategies for estimating the full shift average concentration of a pollutant which is experienced by each of the sampled employees on each trial of an air sampling programme. These are:

(A) Obtain a single measurement of time averaged concentration by operating an integrative personal sampling device for the full period of the shift (of, say, 8 h).

(B) Obtain n measurements of time averaged concentrations by using two or more consecutive samplers covering the whole of the period of the shift. Calculate the weighted mean of the n results from,

$$\bar{x}_w = \frac{\Sigma x_i t_i}{\Sigma t_i}$$

where t_i is the averaging time of the ith air sample.

(C) Obtain n measurements of concentration by using short term or snap samplers at different times within the period of the shift. Calculate the arithmetic mean (or geometric mean) of the n results. The measurements may be made by either simple random sampling, stratified random sampling or systematic sampling, depending upon the exposure pattern.

The choice between these approaches may depend upon a number of situational and other factors, such as the type of work being done by the exposed person, the type of sampling equipment available and costs. From the statistical point of view, approaches A and B can be expected to provide more reliable estimates of the actual shift average concentration than approach C, and thus are preferred. The magnitude of the difference in reliability depends upon the within-shift variability, which in some cases might be fairly small. Where there is considerable within-shift variation, the difference in reliability can be very wide. This is demonstrated by the example in Table 6.8 which compares the precision of two estimates of the 8-h time averaged concentration of styrene experienced by the person depicted in Fig. 6.3. The two estimates were obtained by extracting data from the continuous exposure record in Fig. 6.3 in such a way as to represent the following strategies:

(i) *Strategy B* – four consecutive 2-h air samples. For the purposes of calculating the 95% confidence interval for the true mean, assume the standard deviation of the sampling and analytical technique that might have been used for this strategy to be $\sigma_e = 8$.

(ii) *Strategy C* – fifteen random 3-min samples. Assume that $\sigma_e = 5$.

In this example the precision achieved by strategy B, which is expressed by the

95% confidence interval for the true 8-h concentration, is shown to be far greater than that of strategy C. The effect, under strategy C, of increasing the number of random measurements (that is the sample size, n) on the estimated standard error, $S_{\bar{x}}$, of the 8-h mean concentration, and hence upon the 95% confidence interval for the true 8-h mean, has already been demonstrated in Section 5.3. It can be deduced from Fig. 6.17 that in order to achieve in this styrene example, the same degree of precision by the use of random 3-min air samples as that achieved by strategy B, more than 100 measurements would be required.

8.1.1 Statistical sampling methods

It is stated above that measurements under strategy C can be made by either simple random sampling, stratified sampling or systematic sampling, depending upon the pattern of exposure over the workshift. Simple random sampling, that is making n measurements at random intervals during the workshift, can be used under all circumstances. However, this approach may not always be the most efficient one in the sense of giving maximum precision of estimation of the true 8-h time averaged exposure for a given number of measurements. If, for example, the nature of a person's work is such that his full shift can be divided into time periods (or strata) during each of which there is a fairly steady level of exposure, but between which there are quite distinct differences in exposure level, then stratified sampling would be more efficient.

Stratified sampling in the present context consists of sharing the total number of measurements among the strata so that the number in each is proportional to the length of time involved, the measurements within each stratum being made in a random manner. Suppose that a person's work over 8 h can be divided into two phases or strata, one lasting 3 h, the other 4 h, during which there are quite different levels of exposure, and suppose that the averaging time of the air sampling technique to be used is 5 min. The maximum numbers of consecutive 5-min measurements that

Table 6.8 Comparison of strategies B and C for estimating a person's 8-h average exposure to styrene

Sampling strategy	*Results: Styrene-in-air concentrations (p.p.m.)*			*Mean concentration* $\bar{x}$ *(p.p.m.)*	*Standard error of mean* $S_{\bar{x}}$	*95% confidence interval for mean*
B		4.9		68.9	4	61.06–76.74
		129.4				[Using (6.79)]
		90.8				
		50.4				
C	13.4	1.0	214.8	51.4	22.16	3.87–98.93
	166.3	247.1	0.1			[Using (6.78)]
	10.7	2.0	83.3			
	7.4	0.1	11.9			
	0.1	12.1	0.1			

can be made are $N_1 = 36$ and $N_2 = 60$ respectively, so that the number of measurements made in a stratified sample under strategy C would be $n_1 = N_1/b$ and $n_2 = N_2/b$, where b is a constant. When such stratification of the total assessment period is possible, this sampling method eliminates a source of error and provides an appropriate weighted value of the sample mean, $\bar{x}$, and of the standard error of the mean, $S_{\bar{x}}$. Simple random sampling under the same circumstances might place undue emphasis on one phase or stratum at the expense of another, so that a source of sampling error would be introduced. The standard error of the sample mean would be higher, therefore, than that of the corresponding stratified sample of the same size.

Systematic sampling, that is making measurements at equal time intervals over the full assessment period, can be used in circumstances where there is no systematic variation of pollutant concentration over that time but also where stratification is possible. In the former circumstances the systematic sample would be the same, effectively, as a simple random sample, and under the latter circumstances the systematic sample would have the same advantages in practice as the stratified sample. Neither systematic nor stratified sampling are appropriate methods to use where there is relatively high frequency cyclic variation in pollutant concentrations.

The foregoing discussion also applies to the problem of estimating, for epidemiological purposes, the average exposure level of a population of employees over a much longer assessment period (say, one year) by measuring 8-h time averaged exposures of a selection of them for a number of shifts (trials) within that period. There are two aspects of sampling to which one or other of the three methods can be applied in this case. One aspect is the selection of employees from the population to wear air sampling equipment on each trial (e.g. use of the grouping principles described earlier would lead to stratified sampling) and the other aspect is the timing of the trials. On the question of timing, systematic sampling (i.e. making trials – full shift exposure measurements – at equal time intervals over the year) is likely to be the most convenient approach to use. Due care is necessary in this case to ensure that the sampling trials do not coincide with systematic changes in pollutant concentrations.

Detailed treatment of the three methods of statistical sampling that have been described here can be found in specialised texts on sampling theory.[8,19]

8.1.2 *Frequency of sampling trials*

We now turn to consider the problem of deciding upon the frequency at which trials should be made in a long term air sampling programme. All that is attempted here is to present some statistical decision principles which might be applied in the process of developing an effective and efficient programme.

It may be convenient to plan long term air sampling programmes on an annual basis, that is to consider one year as the assessment period for the purposes of design. Bearing in mind that the basic element of exposure information (the 'property' being measured) here is the 8-h time averaged exposure of an employee, we thus have a potential, finite population of N, 8-h time averaged exposures, where N is given by the number of shifts during the year times the number of employees in the group to be studied. The problem of deciding upon the frequency of sampling trials can now be expressed as one of deciding what annual sample size, n, is sufficient to obtain good estimates of the parameters of the finite population; the sample size being expressed as,

$$n = rk \tag{6.80}$$

where r is the number of persons selected to wear samplers on each trial, and k is the number of trials made over a 1-yr period.

For some purposes, the parameter of main interest might be the annual arithmetic mean pollutant exposure level, $\overline{X}$, which has a normal sampling distribution as demonstrated in Section 5.3. For other compliance monitoring or epidemiological purposes, a formal probability model might be applied to the population distribution and the parameters of interest in this case might be both the mean and the variance of that distribution. For example, the two-parameter lognormal model might be assumed to apply, in which case the parameters to be estimated might be either the mean, θ, and the variance, β^2, or the geometric mean, μ_g, and the geometric standard deviation, σ_g.

Whatever the purpose and whatever distribution model is applied, the planning and development of a long term air sampling programme – including in particular the selection of the annual sample size, n, that will ensure good estimates of the parameters of interest and of certain properties of the distribution – requires provisional estimates to be made of the mean and variance of full shift exposures. It is likely that such estimates would have to be obtained by undertaking a pilot survey of exposures, that is by taking a pilot sample from the population distribution.

The pilot survey would need to be designed to ensure that the two components of the total variance of 8-h time averaged exposures are adequately covered. One of these components is the variance between members of the study group, within shifts; the other is the between-shift variance which, in general, is considerably larger. The total variation in the case of a lognormal population distribution can be expressed in terms of the geometric standard deviation, as follows:

$$\sigma_g = \exp\{(\sigma_W^2 + \sigma_B^2)^{0.5}\}$$

where σ_W^2 and σ_B^2 are respectively the within-shift and between shift variances of logarithmic values of 8-h time averaged exposures.

The pilot survey would need to extend, therefore, over a number of shifts and to cover periods of any known, systematic shift-to-shift variation in processes and methods of work which are typical of long term operations. It would also need to be designed to facilitate the grouping of employees according to the similarity principles that were described earlier in this section.

Having obtained provisional estimates of the mean and variance of 8-h exposure levels from a pilot sample (of size n_0) a tolerance can be set on the precision of estimation, by reference to the $100(1-\alpha)\%$ confidence interval for the true value of one or other of these parameters. Thus we might specify that either the upper or the lower limit of the $100(1-\alpha)\%$ confidence interval must be within $P\%$ of the estimate of the parameter. We can then calculate the annual sample size, n, that is required to achieve the specified tolerance, the difference $n-n_0$ being the amount by which the pilot sample size has to be increased. Two examples of expressions that can be so derived for calculating n are given below.

Firstly, if the chief interest is to estimate the annual arithmetic mean exposure level, $\overline{X}$, for a group of employees, then by use of Expression 6.23 we can specify that

$$\text{either } \bar{x}-\left(\bar{x}-t_{\alpha/2,\nu}\cdot S\sqrt{\frac{N-n}{n(N-1)}}\right)\leq P/100\,\bar{x}$$

$$\text{or } \left(\bar{x}+t_{\alpha/2,\nu}\cdot S\sqrt{\frac{N-n}{n(N-1)}}\right)-\bar{x}\leq P/100\,\bar{x}$$

From either of these specifications we obtain that

$$n\geq N\left\{(N-1)\left(\frac{P/100\,\bar{x}}{t_{\alpha/2,\nu}\cdot S}\right)^{2}+1\right\}^{-1} \tag{6.81}$$

where N is the population size, P% is the specified tolerance, $\bar{x}$ and S are the mean and standard deviation of the pilot sample of size n_0, and where $t_{\alpha/2,\nu}$ is the value of the t-distribution corresponding to the desired confidence level and to $\nu=n_0-1$ degrees of freedom. The number of trials, that is the number of shifts to be sampled over the year, is then given by $k=n/r$.

Secondly, if it is required to estimate the parameters of the two-parameter lognormal distribution on the assumption – or perhaps in the knowledge – that this model adequately describes the population distribution, then we can use the upper limit of the $100(1-\alpha)$% confidence interval for the geometric mean, given by Expression 6.53, to specify that

$$M_g \exp\left\{t_{\alpha/2,\nu}\cdot \ln S_g\sqrt{\frac{N-n}{n(N-1)}}\right\}-M_g\leq P/100\,M_g$$

From this specification we obtain that

$$n\geq N\left\{\frac{\ln^2(1+P/100)(N-1)}{(t_{\alpha/2,\nu}\cdot \ln S_g)^2}+1\right\}^{-1} \tag{6.82}$$

where S_g is the geometric standard deviation of the pilot sample and where the other terms are as defined for Equation 6.81.

Further adjustment of the annual sample size, that is of the number of trials and the number of persons involved in each trial, may need to be made as the air sampling programme progresses and more data are accumulated. Whenever there are major alterations in the processes carried on in the working environment, in the materials used, or in provisions for the control of emissions of the pollutant gas or vapour, it is important to review the overall strategy of the air sampling programme to decide whether or not there is a need to modify it in some way.

In the case of monitoring compliance with an 8-h exposure limit such as a Threshold Limit Value (TLV) recommended by the American Conference of

Governmental Industrial Hygienists, it can be argued that the frequency of air sampling trials should depend upon how far employees' 8-h time averaged exposures differ from the TLV for the gas or vapour in question. Roach[42] has suggested the minimum frequencies given in Table 6.9. Different frequencies are suggested for different mean values of the ratio of 8-h personal exposure to the TLV. Roach does not appear to take account, here, of within-shift and between-shift variances in personal exposures.

8.2 Area Monitoring

Area monitoring by static sampling is done mainly for the purpose of surveillance of the long term performance of engineering provisions for the control of emissions of a pollutant gas or vapour from plant and processes. The siting of static sampling points for this purpose is determined by the existing and potential sources of emission of the pollutant. Due account should be taken of the patterns of airflow through the work area and around the sources of emission – bearing in mind that the general ventilation and aerodynamics of the work area may be subject to seasonal variation – when deciding upon the location and number of sampling points.

Much of the discussion in the previous section is relevant to the problem of formulating a strategy for area monitoring. The required frequency and averaging time of sampling depends upon the within-shift and between-shift variation in pollutant concentrations. Where a target value of, say, the annual arithmetic mean concentration of the pollutant is set for the work area, for statistical control or other purposes, the frequency of sampling also depends upon the desired precision of estimation of the actual mean value. With regard to the averaging time, it may be found that shift (8-h) averaging is suitable in many circumstances, but longer averaging times may be required sometimes in order to detect significant trends.

The use of a statistical control chart[43–45] for the analysis and display of air sampling results may be of considerable value as a means of detecting any significant deterioration in control and indicating when corrective action is required.

Table 6.9 The minimum time to be spent on regular monitoring personal exposure – suggested by Roach[42] (from *Annals of Occ. Hygiene,* Vol. 20, 65–84, 1977)

Man-shifts covered by personal sampling (per 10 employees)[1]	*Ratio of personal exposure to 8-h time weighted average threshold limit value*
1/month	1–2
1/quarter	0.5–1 or 2–4
1/annum	0.1–0.5 or 4–20
None[2]	<0.1 or >20

[1] If there are less than 10 employees in the job under surveillance, assume 10.

[2] Provided that there has been no change of process, material or contaminant control since the last survey.

References

1. J. E. Peterson, H. R. Hoyle and E. J. Schneider. The application of computer science to industrial hygiene. *Amer. Ind. Hyg. Assoc. J.* **27**, 180–5 (Mar.–April 1966).
2. V. Dharmarajan, H. Weil and C. W. Self. Environmental characterisation of toluene diisocyanate (TDI) in a manufacturing plant. *Amer. Ind. Hyg. Assoc. J.* **39** (May 1978).
3. W. Weibull. A statistical distribution function of wide applicability. *J. Applied Mech.* **18**, 293–7 (1951).
4. Hald, A. Statistical Theory with Engineering Applications. Wiley (1952).
5. M. G. Kendall and A. Stuart. *The advanced theory of statistics,* 3rd revised edn. Vol. 1, *Distribution Theory*; Vol. 2, *Inference and Relationships.* Charles Griffin (vol. 1: 1977; vol. 2: 1973).
6. O. M. Essenwanger. *Applied Statistics in Atmospheric Science 4A. Part A: Frequencies and Curve Fitting.* Elsevier Scientific (1976).
7. J. Aitchison and J. A. C. Brown. *The Lognormal Distribution.* Cambridge University Press (1957).
8. V. Barnett. *Elements of Sampling Theory.* The English Universities Press (1974).
9. W. G. Cochran. *Sampling Techniques,* 2nd edn. Wiley (1963).
10. J. Murdock and J. A. Barnes. *Statistical Tables for Science, Engineering and Management.* Macmillan (1968).
11. N. A. Esmen and Y. Y. Hammad. Log-normality of environmental sampling data. *J. Environ. Sci. Health,* **A12** (1 and 2), 29–41 (1977).
12. D. J. Finney. On the distribution of a variate whose logarithm is normally distributed. *J. Roy. Statist. Soc., Suppl.* **7**, 155–61 (1941).
13. N. A. Rahman. Application of the χ^2 distribution. *A Course in Theoretical Statistics,* Chap. 17. Charles Griffin (1968).
14. F. J. Massey. The Kolmogorov–Smirnov test for goodness of fit. *Amer. Statistical Assoc. J.* **46**, 68–73 (1951).
15. S. Siegel. Kolmogorov–Smirnov test. *Nonparametric Statistics for the Behavioural Sciences,* pp. 47–52. McGraw-Hill (1956).
16. N. A. Leidel, K. A. Busch and J. R. Lynch. *Occupational Exposure Sampling Strategy Manual.* DHEW (NIOSH) Publication No. 77–173, Cincinnati, Ohio (1977).
17. Y. Bar-Shalom, A. Segall and D. Budenaers. Decision and estimation procedures for air contaminants. *Amer. Ind. Hyg. Assoc. J.* **37**, 469–73 (1976).
18. R. S. Brief and A. R. Jones. A statistical technique for determining compliance with dual hygienic standards. *Amer. Ind. Hyg. Assoc. J.* **37**, 474–8 (1976).
19. R. J. Sherwood. The monitoring of benzene exposure by air sampling. *Amer. Ind. Hyg. Assoc. J.* **32**, 840–6 (1971).
20. W. Coenen. On the problem of the confidence region for average values of dust concentration. *Staub-Reinhalt, Luft.* **26** (5), 39–45 (1966).
21. R. J. Sherwood. On the interpretation of air sampling for radioactive particles. *Amer. Ind. Hyg. Assoc. J.* **27**, 98–109 (1966).
22. H. J. Gale. Some examples of the application of the lognormal distribution in radiation protection. *Ann. Occup. Hyg.* **10**, 39–45 (1967).
23. R. F. Hounam. *An Application of the Lognormal Distribution to some Air Sampling Results and Recommendations on the Interpretation of Air Sampling Data.* Atomic Energy Res. Establishment Rep. AERE-M 1469, HMSO, London (1965).
24. H. Gormar. Beitrag zum Problem der Statistik bei der Raumluftuberwachung auf einen Schadstoff am Beispiel Vinylchlorid. *Staub-Reinhalt. Luft,* **36** (4), 160–8 (1976).
25. R. I. Larsen. A new mathematical model of air pollutant concentration, averaging time and frequency. *J. Air Pollution Control Assoc.* **19**, 24–30 (1969).
26. R. I. Larsen *et al.* Analyzing air pollutant concentration and dosage data. *J. Air Pollution Control Assoc.* **17**, 85–93 (1967).

27. R. I. Larsen. An air quality data analysis system for interrelating effects, standards, and needed source reductions. *J. Air Pollution Control Assoc.* **23**, 933 (1973).
28. N. D. Singpurwalla. Extreme values of a lognormal law with applications to air pollutant problems. *Technometrics,* **14**, 703–11 (1972).
29. G. Berry and N. E. Day. The statistical analysis of the results of sampling an environment for a contaminant when most samples contain an undetectable level. *Amer. J. Epidemiology,* **97**, 160–6 (173).
30. Y. Kalpasanov and G. Kurchatova. A study of the statistical distribution of chemical pollutants in air. *J. Air Pollution Control Assoc.* **26**(10), 981–4 (1976).
31. T. C. Curran and N. H. Frank. Assessing the validity of the lognormal model when predicting maximum air pollutant concentrations. Paper No. 75-51.3. *68th Annual Meeting of the Air Pollution Control Association* (1975).
32. J. A. Tikvart and W. P. Freas. A frequency distribution analysis of calculated pollutant concentrations, infrequent meteorological conditions and allowable emissions. *Fifth Conference on Probability and Statistics in Atmospheric Sciences,* Nov. 15–18, Las Vegas, NV (1977).
33. W. R. Ott and D. T. Mage. A general purpose univariate probability model for environmental data analysis. *Comput. and Ops. Res.* **3**, 209–16 (1976).
34. D. T. Mage and W. R. Ott. Refinements of the lognormal probability model for analysis of aerometric data. *J. Air Pollution Control Assoc.* **28**(8), 796–8 (1978).
35. B. E. Saltzman. Significance of sampling time in air monitoring. *J. Air Pollution Control Assoc.* **20**(10), 660–5 (1970).
36. O. L. Davies (ed.). *The Design and Analysis of Industrial Experiments,* 2nd edn. Longman (1978).
37. B. J. Winer. *Statistical Principles in Experimental Design.* McGraw-Hill (1962).
38. N. R. Draper and H. Smith. *Applied Regression Analysis.* Wiley (1966).
39. *Threshold Limit Values of Chemical Substances and Physical Agents in the Workroom Environment with Intended Changes for 1979.* American Conference of Governmental Industrial Hygienists, PO Box 1937, Cincinnati, Ohio (1979).
40. German Research Society (Deutsche Forschungsgemeinschaft). *Maximum Concentrations at the Workplace,* Report No. 15 of the Commission for Investigation of Health Hazards of Chemical Compounds in the Work Area, 5 June 1979.
41. M. Corn and N. A. Esmen. Workplace exposure zones for classification of employee exposures to physical and chemical agents. *Amer. Ind. Hyg. Assoc. J.* **40**, 47–57 (1979).
42. S. A. Roach. A most rational basis for air sampling programmes. *Ann. Occ. Hyg.* **20**, 65–84 (1977).
43. J. Morrison. The lognormal distribution in quality control. *Applied Statistics,* **7**(3), 160–72 (1958).
44. C. A. Bennett and N. L. Franklin. *Statistical Analysis in Chemistry and the Chemical Industry.* Wiley (1954).
45. O. L. Davies and P. L. Goldsmith. *Statistical Methods in Research and Production: With Particular Reference to the Chemical Industry,* 4th revised edn. Longmans (1976).

Appendix

Glossary of symbols

Parameters of population distributions:

- μ — mean of normal distribution of a variate
- σ^2 — variance of normal distribution
- σ — standard deviation of normal distribution
- θ — mean of lognormal distribution
- β^2 — variance of lognormal distribution
- μ_g — geometric mean or median of lognormal distribution
- σ_g — geometric standard deviation of lognormal distribution
- τ — displacement parameter, lognormal distribution
- $Q_{100(1-\alpha)}$ — value of a variate corresponding to the $100(1-\alpha)$ percentage point or quantile of its distribution

Sampling distribution statistics:

- $\bar{x}$, $\bar{y}$ — arithmetic mean values of variates X and Y
- S^2 — sample variance
- S — sample standard deviation
- S_a — sample standard deviation, lognormal distribution
- S_g — geometric standard deviation
- M_g — geometric mean
- M_a — sample mean, lognormal distribution
- c — displacement parameter, lognormal distribution
- $q_{100(1-\alpha)}$ — value of variate corresponding to the $100(1-\alpha)$ percentage point or quantile (estimated)

Other symbols:

- n — number of elements in a statistical sample
- N — number of elements in a finite population
- T — assessment period over which the distribution of pollutant concentrations in a particular situation is to be described
- t — sampling or averaging time.

7 Standard atmospheres

A. F. Smith

1 Introduction

An essential requirement for the development of tests for the detection and estimation of hazardous gases and vapours which may be present in the atmosphere, and also for the calibration of instruments used for this purpose is the provision of accurate standards. Chemical standards are usually prepared by dissolving given amounts of a substance of known purity in a suitable substrate; gases and vapours are no exception. However, although it is a relatively easy matter to prepare an aqueous solution of, for example, potassium dichromate which is a fairly easily weighed solid of high purity, standard solutions of gases and vapours in air can be difficult to prepare and special techniques are required. Standard mixtures of gases and vapours are generally referred to as standard atmospheres although they should perhaps be more correctly called mixtures of known composition since it is not always possible to reproduce exactly the same concentration even when using the same technique.

There is no universally applicable method of preparing standard gas mixtures as the technique to be used depends on many factors not least of which is the purpose for which the standard is required. Other factors which may influence the choice of method include the complexity of the mixture, its concentration, the chemical reactivity of the constituents, sorption of the constituents by the materials used in the apparatus and the accuracy required of the final product.

There are two general methods of preparation: static and dynamic. The static method involves the introduction of known quantities of gases or vapours into suitable vessels of known volume containing the diluent gas and allowing them to mix to form the desired standard. In the dynamic method, gases or vapours are introduced at measured rates into a flowing air or diluent gas stream to give a continuous supply of the desired standard.

The general principles and techniques of preparation of gas mixtures are described in some detail by Nelson[1] and are the subject of a British Standard.[2] They have also been reviewed by many other authors and there is an extensive literature dealing with the preparation of specific substances. It is not intended that this chapter should provide working instructions for the preparation of standard atmospheres, but that it should attempt to summarise some of the principles involved, the limitations of various methods and the main pitfalls to avoid.

2 General considerations

The generation of standard atmospheres requires a working knowledge of the fundamental laws relating to the behaviour of gases under varying temperature and pressure conditions. For most practical purposes the majority of gases and vapours obey the ideal gas laws at room temperature and atmospheric pressure, but for the highest accuracy any deviations must be taken into account, especially at high pressures. In the field of air pollution, gas and vapour concentrations are usually expressed as parts per million by volume, but practical considerations and the analytical method that may be used for checking the generated atmosphere concentration, if this is necessary, may require calculations involving concentrations expressed on a weight per unit volume basis, e.g. $mg\,m^{-3}$. Ability to convert from one measurement system to another is therefore also required.

3 Apparatus

Glass is the preferred material for use in the construction of apparatus for generating standard gas mixtures, although more flexible materials such as plastics may be more convenient for static systems. The general lack of flexibility of glass is outweighed by the fact that it is easily worked and for most substances adsorption losses are generally low. A wide range of commercial fittings is also available. All-glass systems are highly desirable and this can be achieved by the use of ground, ball-and-socket or standard taper joints. The former are to be preferred since they impart a measure of flexibility to the apparatus. Unless high back pressures are developed within the apparatus, and this should not be the case in a well-designed generator, ball-and-socket joints fastened with spring clips are sufficiently gas tight. Back pressures within the apparatus can be minimised by using tubing with the widest possible bore and by avoiding sharp bends and unnecessary constrictions. The use of grease on glass joints should be avoided if possible to remove another potential source of sorption from which the vapour may later be released. However, some components of a gas mixture may have a high diffusion coefficient and greased joints may be necessary to ensure complete gas-tightness; if this is the case, then the grease should be inert to the components of the mixture and the amount used should be as small as possible. Sometimes, the use of glass is not possible, e.g. for the preparation of hydrogen fluoride with which it will react. In this case, polythene has been used successfully.[3]

4 Flow measurement

The production of gas mixtures requires an accurate control of the volume and flow-rate of the constituents if a high degree of accuracy is to be achieved. Gas flow may be measured by many methods but those most commonly used are variable-area direct reading flowmeters, capillary flowmeters, bubble flowmeters and wet or dry gas meters.

4.1 Variable-area flowmeters

These meters are probably the most widely used instruments for measuring flow-rates when preparing gas mixtures; they are available commercially to cover a wide range of flow-rates and can be supplied calibrated for the gas of choice. The meters take the form of a tube, increasing in diameter from bottom to top, in which a float is free to move. The gas enters the tube at the bottom and the float is raised thereby increasing the annular area between the float and the tube until the upward forces are balanced by the weight of the float. The meters are direct-reading instruments usually calibrated for air at atmospheric pressure and room temperature, but suppliers will calibrate, on request, at any of a wide range of pressures and temperatures for most gases. Individually calibrated flowmeters of this type have floats which are not interchangeable. Some manufacturers supply flowmeters with an uncalibrated, engraved scale and a set of interchangeable floats to cover a range of flow-rates; such meters require calibration by the user.

Variable-area flowmeters will of course only give a reading within the calibration accuracy, usually of the order of $\pm 2\%$, when operated under the pressure and temperature conditions specified by the manufacturer. Under the closed conditions of an atmosphere generator, any back pressures induced will affect the calibration. The correct flow-rate may be calculated, but it is usually more convenient, in practice, to accept the reading of the flowmeter as being approximate. In this case standardisation of the generated atmosphere must be carried out. Most variable-area flowmeters are calibrated at atmospheric pressure and room temperature and it is sometimes possible to introduce, temporarily, into the circuit, a second meter open to atmosphere to give an indication of the true flow.

4.2 Capillary flowmeters

These meters are generally more suited to the measurement of low flow rates. They consist essentially of a restricting orifice, usually a capillary tube, placed in the moving gas stream. The pressure differential produced is proportional to the flow rate and is measured by a manometer connected across the tube. The meter may be calibrated against a bubble flowmeter or a wet or a dry gas meter. Capillary flowmeters are particularly useful where the viscosity and density of the gas being measured make the use of conventional flowmeters impracticable.

4.3 Bubble flowmeters

In this type of meter[4] the gas is passed through a bubble-forming solution, usually a dilute solution of a surfactant, and the bubbles are carried up a calibrated tube. The flow rate is measured by timing the rise of one of these bubbles between two selected volume markings. The use of a bubble flowmeter is limited to gases which are not soluble in the liquid used to generate the bubbles. The accuracy of the method is generally of the order of ±1% but considerable deviations from this can occur depending on the design of the meter[5] and, in the case of low flow rates, the gas being measured.[6] The use of electrical timing devices[7,8] rather than a stop-watch can also improve accuracy. Bubble flowmeters are rarely used as an integral part of an atmosphere generator, but are very useful for calibrating variable-area and capillary meters, restrictors and other flow measuring or control instruments. Practical difficulties restrict the use of the method to the measurement of flow rates below about 500 cm^3 min^{-1}.

4.4 Wet gas meters

A wet gas meter consists of a container housing a revolving drum partly submerged in water. The drum has a number of compartments each with an inlet and an outlet port. The force of the gas entering a compartment causes the drum to rotate until the next compartment reaches the filling position and the full compartment expels its gas through the outlet. The drum therefore rotates smoothly as gas is passed into the meter at a steady rate and the total volume for a given time is recorded by suitably geared dials. The range of flow rates over which this type of meter may be used is about 1–50 dm^3 min^{-1} since at lower flow rates the friction losses are high and at higher rates excessive pressure differentials may occur. For the highest accuracy the gas being metered should be saturated with water vapour at the temperature of the meter and the water itself saturated with the gas. This may mean passing gas through the meter for several hours to ensure that equilibrium is reached. Wet gas meters are accurate to within ±1% of the indicated values provided that the necessary precautions are taken to ensure that errors are minimal.

4.5 Dry gas meters

The dry gas meter, similar to that used for metering domestic, natural gas supplies, provides a useful means of measuring flow rates in the range of about 5–5000 dm^3 min^{-1} accurate to about ±2%. The passage of gas through these meters alternately inflates and deflates bellows which actuate the flow indicating mechanism. Unlike wet meters there is no problem of vapour equilibration, but care must be taken to ensure that there is no absorption of the gas by the materials used in their construction.

Both wet and dry gas meters should be checked periodically against a primary standard such as a spirometer. As with bubble flowmeters the main use of these meters is for the calibration of other flow measuring instruments, but they can be used for direct measurement of gas flows although their use as an integral part of an atmosphere generator is rare.

4.6 Mass flowmeters

A number of techniques are now available for measuring flow rates covering a range from less than 0.1 cm^3 to several thousand dm^3 per minute. These are known collectively as mass flowmeters and the majority are based on heat transfer methods.[9,10] A thin, heated wire or a thermistor placed in a moving gas stream will lose heat depending on the velocity and density of the gas, and the change in temperature of the wire results in a proportionate change in resistance which can be measured by means of a bridge circuit. These instruments, which are expensive, can be very accurate but require calibration against a primary standard. Their major disadvantage is that the presence of heated wires makes them unsuitable for use in explosive or inflammable atmospheres and with reactive or corrosive gases.

5 Flow control

A reliable and accurate means of controlling the flow of gases at a steady rate is an essential component of any standard atmosphere generator. Parts of a system will require a fixed flow rate whereas in other parts, particularly with dynamic systems, flexible control will be desirable. The most widely used methods of control at a constant flow rate are probably the critical orifice[11–18] and the restrictor, both of which can be inexpensive, convenient and easy to install and operate. They are, however, susceptible to damage and blockage, but with reasonable care and by taking the necessary precautions to remove particulate matter from the gas stream these disadvantages can be alleviated. Sections of glass or metal tubing, capillary tubing,[16,19] watch jewels,[13,17] hypodermic needles,[14,15,20] flattened metal tubes[21] and sintered metal or glass plugs have all been used as restricting orifices. The material used will depend on the nature of the gas that it is required to control. The flow rate of a gas through a restrictor depends on the pressure difference between the inlet and the outlet. An orifice or restrictor is said to be critical when the ratio of the downstream pressure (P_D) to the upstream pressure (P_U) produces a sonic velocity at the orifice outlet.[22] For air, this condition is met when P_D is less than 0.53 P_U and the ratio of the cross-sectional areas of the tube carrying the gas to the orifice and the orifice itself is greater than 25.[11,22] At the critical state further increase in the pressure differential will have no effect on the flow-rate. Flow control by this means is very convenient in dynamic systems where the primary atmosphere is obtained by the direct dilution method (p. 204).

Adjustable valves provide a convenient means of controlling and varying the flow rate of gases. These may range from simple 'on–off'-type valves suitable for crude control of high flow rates of the order of 10–100 $dm^3\ min^{-1}$ to expensive, sensitive needle valves capable of high precision control of rates ranging from a few cm^3 to 10+ $dm^3\ min^{-1}$.

For extremely low flow rates a porous plug flowmeter[23,24] may be used. The gas to be controlled is split into two streams, one of which passes through a capillary tube packed with asbestos fibres and the other to waste through a column of liquid. The flow rate obtained will depend on the tightness of packing of the plug and the height of the liquid column in the waste line.

Flow-indicating instruments may sometimes show variations in flow rate during use. These are often due to fluctuating gas pressures or to flow patterns within the

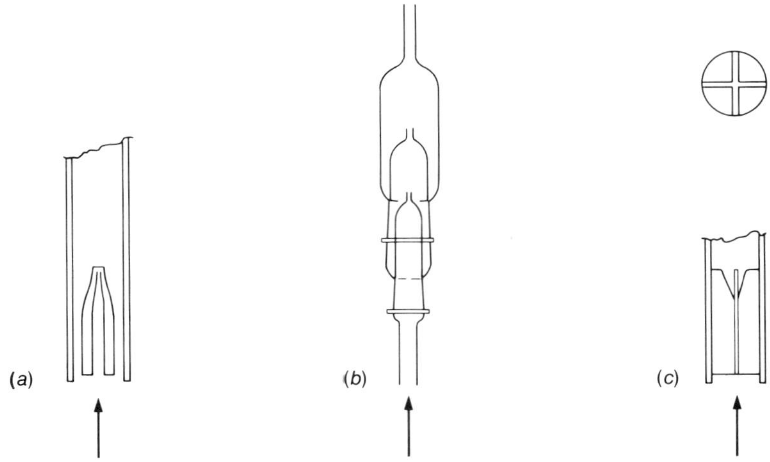

Fig. 7.1 Some devices for reducing fluctuations in gas flow rates. (*a*) Flow smoother; (*b*) ripple smoother; (*c*) flow straightener.

apparatus and may be reduced or eliminated by simple devices placed at suitable points in the gas stream. A constriction in the form of a coarse jet (Fig. 7.1 *a*) or of a ripple smoother (Fig. 7.1 *b*)[25] immediately before the flowmeter is effective in damping out fluctuations caused by pressure changes in the gas supply. Rapid or short-term pressure fluctuations may also be overcome by the use of an automatic pressure regulator. This type of regulator is particularly useful when it is necessary to supply a number of independently controlled dilution stages from one source of compressed gas. Sharp bends in tubing leading to flowmeters may give rise to uneven flow patterns causing instability of the flowmeter, and these irregularities may be reduced by insertion of a flow straightener, in the form of a series of parallel channels, into the inlet tube of the meter. A simple and effective flow straightener (Fig. 7.1 *c*) can be made from two thin rectangular strips of metal, plastics or card fitted into the tube diametrically and at right angles to one another and end-on to the direction of the gas flow. In cases of extreme turbulence more than two strips may be necessary to achieve stability.

6 The preparation of gas mixtures by static methods

Despite their limitations, static methods for the preparation of gas mixtures have the advantage of being relatively simple and provided that only small volumes of the standard are required over short periods of time as in the calibration of single instruments, e.g. gas chromatographs and gas detector tubes, they can be very useful. One of the most convenient methods of making a gas standard is to introduce a predetermined amount of solvent or gas into a single, rigid container of known volume containing the diluent gas. Vaporisation of a solvent may be necessary before dilution. The containers may take many forms from small syringes,[26,27] bottles and flasks[28–32] to specially fabricated test chambers and rooms.[33–36] To ensure that the system is kept at atmospheric pressure it is necessary to provide an inlet for replacement gas when a sample is withdrawn for use. For this reason the size of the vessel must be fairly large, usually about 20–40 litres,[28–30] if excessive dilution of the standard is to be avoided. The materials used for construction of the apparatus should be smooth and such that sorption by, reaction with and permeability to the components of the mixture is low. Glass is most commonly used, but plastics[29–31,37] and metals[34,38,39] have also been used. A comparison of materials suitable for use has been made by Wilson and Buchberg.[40] Introduction of the contaminant gas or liquid into the vessel can be accomplished in a number of ways, the simplest method being to inject by means of a calibrated syringe through a thin septum built into the vessel.[41–43] Another method is to break glass ampoules containing known amounts of the pure contaminant within the dilution vessel.[29,41] The contaminant gas can be produced *in situ* by reacting stoicheiometric amounts of reacting chemicals. Hydrogen cyanide atmospheres have been produced in this way by reaction of potassium cyanide with concentrated sulphuric acid.[1] For relatively high concentrations, e.g. within the percent rather than p.p.m. range, gas blenders using volumetric gas burettes may be used.[43–46]

By their very nature many static systems require means to ensure that the constituents are properly mixed. The method adopted will of course depend on the physical properties of the components. The most common means and one of general application is some form of stirrer. For small systems a simple magnetic stirrer will suffice,[43] but generally impellers, some of which may be of ingenious design driven by external motors, are necessary. For very large systems which may approach the size of small rooms, circulation fans have been used.[33,34,37] Additionally, in cases where the contaminant is introduced in liquid form and particularly where the liquid has a high boiling point, some means of vaporisation such as a heating lamp[47] or mantle will be necessary. Dispersion of the liquid on to adsorbent paper[29] or cotton wool increases the surface area and helps to speed evaporation. For obvious safety reasons, particularly when dealing with inflammable substances, adequate precautions must be taken to ensure that devices used for mixing and/or heating are not likely to initiate fire or explosion.

Static systems have a number of disadvantages not least the fact that sorption and reaction may occur on the walls of the container.[48–50] Losses due to sorption may be as high as 50%[40] but the effect can be reduced by conditioning the system before use. This requires allowing the gas mixture to remain in the apparatus for some time, then emptying the system and repeating the process several times before the prepared mixture can be used with confidence as to its concentration. The procedure is tedious and time-consuming but if a static system must be used then for the greatest accuracy it must be followed. Sorption losses may be reduced by increasing the size of the container thereby reducing the surface to volume ratio, but even in large vessels significant reductions in concentration in a relatively short time have been reported.[37,49] Reaction losses are greatest with strong oxidising and reducing agents[48] and polymerisable materials.

If a partial vacuum is not to be introduced in a rigid container, thereby creating the possibility of an implosion, gas must be introduced to maintain equilibrium when a sample of the gas mixture is withdrawn. If this make-up gas is the diluent gas, then the gas mixture is diluted each time a sample is taken. Assuming instantaneous and perfect mixing of the incoming diluent gas with the remaining gas mixture then the change in concentration is equal to the concentration times the fraction of the volume withdrawn:

$$\mathrm{d}C = \frac{C\mathrm{d}V_s}{V}$$

integrating $C = C_o \mathrm{e}^{-V_s/V}$

or $\quad 2.303 \log_{10} C_o/C = V_s/V$

where C = concentration at any time; C_o = original concentration; V = volume of container; V_s = volume of sample withdrawn.

Dilution by a make-up gas in rigid systems can be overcome in two ways. One method is to attach a deflated, flexible bag to the dilution gas inlet[51]; as the mixture is sampled, dilution gas fills the bag and the effective volume of the container is reduced correspondingly. Care must be taken to ensure that the bag material does not adsorb or absorb the contaminant gas. A second method is to use a number of containers in series, each of which is filled with a mixture of the desired concentration. As a sample is withdrawn it is replaced by an equal volume of gas from the following container and successively along the series.[41] Dilution will of course still occur but only to a small degree and will depend on the number of vessels in the train. One major disadvantage of such a system is the replication of apparatus and the large amount of space required to set it up.

The disadvantage of rigid systems can be overcome by the use of flexible containers, usually bags made of a suitable plastics material. Samples can be withdrawn without replacement with air of gas mixture since the walls of the bag will conform to the reduced volume with negligible change in internal pressure provided that it is not filled to capacity and the internal pressure does not exceed atmospheric. Compared with the apparatus required for rigid systems these bags are cheap and are available commercially; they may also be fabricated in the laboratory. Their lightness and compactness make them particularly suitable for use for instrument calibration in the field. However, plastics bags must be used with care since changes in composition of the stored gas mixture may take place as a result of diffusion through and decomposition and adsorption on the walls, especially during prolonged storage.[40,48,52–56] The permeability of the plastics material may be reduced by bonding it to a metal, usually aluminium, layer.[52]

Gas mixtures in pressurised cylinders are obtainable from commercial manufacturers and in recent years the available range of gases, concentrations and cylinder sizes has increased considerably and many suppliers will produce mixtures to special order. Although gas mixtures in pressurised cylinders may be produced in the laboratory, the control difficulties and the fairly complex apparatus required usually make it more convenient and economic to purchase. Most manufacturers supply a certificate with their mixtures. This gives details of concentration, method and date of production, accuracy and other information that may be considered necessary with regard to storage, safety precautions, etc. The concentration quoted is that at the time of certification and many factors will determine whether this will remain constant throughout the life of the cylinder. There will be, for example, sorption of the gases on the inner surfaces of the cylinder, the amount of which will depend on the nature of the gases and cylinder material, temperature and particularly pressure.[29,50,52] Thus as the mixture is used and the internal pressure falls the concentration may change, sometimes considerably. It is as well therefore to check the concentration of the gas mixture from time to time particularly if a partially used cylinder has been stored for some time.

Space does not permit a detailed description of the production of pressurised gas mixtures. The three main methods,[2] volumetric, manometric or partial pressure and gravimetric, can be used to produce mixtures with concentrations from the low parts per million range to the percent range and Nelson[1] quotes average accuracies of the order of ±2%, ±5% and ±0.5%, respectively, for these methods.

6.1 Calculations

The theoretical concentration of the contaminant gas or vapour in a binary mixture can be calculated as follows:

$$\text{Concentration} = 10^6 \times \frac{V_c}{V_d} \text{ p.p.m. by volume}$$

or

$$\frac{10^2 \times V_c}{V_d + V_c} \text{ per cent by volume.}$$

where V_c = volume of contaminant gas (cm^3); V_d = volume of diluent gas (cm^3).

For a binary mixture produced by the vaporisation of a known weight of liquid contaminant, the concentration can be calculated from the expression

$$\text{Concentration} = \frac{10^6 \times 22.4 \times W}{VM} \cdot \frac{T}{273} \cdot \frac{760}{P} \text{ p.p.m. by volume}$$

where V = volume of system (dm^3); W = weight of liquid (g); M = molecular weight of contaminant (g); T = Temperature of system (°K); P = Pressure of system (mm).

It is often more convenient to add the liquid contaminant by volume rather than weight in which case

$$W = \rho V_c$$

where ρ = density of the liquid ($g\,cm^{-3}$); V_c = volume of contaminant (cm^3).

7 The preparation of gas mixtures by dynamic methods

Dynamic systems for producing gas mixtures have a much wider range of applicability than static systems; more complicated and expensive apparatus may be required but they generally suffer from fewer disadvantages particularly where a continuous supply of large volumes of a mixture is required over a long period of time. Their relatively complex nature makes them generally more suited to the laboratory than the field and although widely used for calibration of gas detection instruments they are particularly useful for test development and for checking the efficiency of sampling methods and techniques. Perhaps the greatest advantage of dynamic over static systems is the reduced effect of adsorption on the surfaces of the apparatus. Once equilibrium has been established such losses are usually negligible provided that the concentration remains constant. Changes in concentration, which are very easily achieved in properly designed dynamic systems, may upset the equilibrium for a short time but stability is attained fairly quickly especially if the apparatus has been in use for some time and is well conditioned. The speed with which a new generator will reach equilibrium depends on the nature of the gas mixture; strongly polar substances may take a considerable time. Atmosphere concentrations covering the percent to the parts per billion range are possible with a single dynamic system although in practice it is not usual for generators to be expected to cover more than a small range of concentrations.

A dynamic system may be divided conveniently into three main parts: the concentrated atmosphere source or primary atmosphere, the dilution stage or stages, and the sampling stage (Fig. 7.2).

7.1 The concentrated atmosphere source or primary atmosphere

The choice of method for producing the primary atmosphere will be determined mainly by the properties of the proposed diluent gas and of the added substance. The gas used for dilution will depend on the purpose for which the atmosphere is intended to be used and so far as calibration of gas detection instruments is concerned this will be almost invariably air. Properties of the added substance to be considered include:

(*a*) physical nature at room temperature which, in conjunction with volatility, determines to a large extent what general method of generation may best be used;

(*b*) vapour density and viscosity, the latter may affect the calibration of flowmeters;

(*c*) adiabatic expansion and cooling which may also affect flowmeters and is often appreciable for liquids with low boiling points;

(*d*) chemical reactivity with materials used in the construction of the generator.

Methods of introducing the contaminant gas into the diluting gas stream have been classified in many ways but may be divided loosely into five groups: direct dilution, injection, vapour saturation, diffusion, and permeation.

7.1.1 Direct dilution methods

This technique is used for gases and vapours obtainable from cylinders either as compressed or liquefied gas, a previously diluted gas or from a suitable chemical or electrolytic gas generator. Commercial gas manufacturers now produce a wide range of high purity gases in cylinders and the need to produce a gas *in situ* by chemical

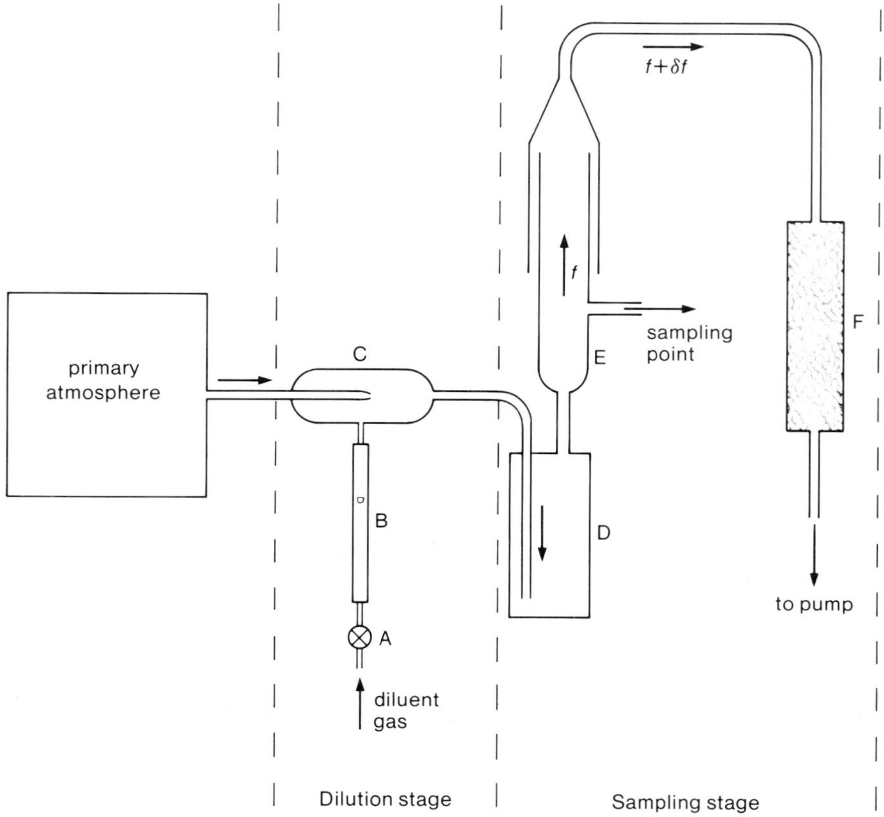

Fig. 7.2 Schematic diagram of apparatus for the preparation of gas mixtures by dynamic methods. A, Flow control valve; B, flowmeter; C, mixing chamber; D, reservoir; E, sample bell; F, waste trap; *f*, flow rate.

means is rarely necessary. To obtain the required concentration with minimum subsequent dilution the flow of gas from the cylinder to the diluting air stream should be kept as low as possible. This limits the type of meter that can be used to measure the rate of flow. Variable-area flowmeters are generally unreliable below about 50 cm^3 min^{-1} and would in any case need to be calibrated for the gas being used; the most suitable means of flow control or measurement is therefore probably a porous plug flowmeter, a capillary flowmeter, a bubble flowmeter or a restricting orifice. With these metering devices it is necessary to maintain a constant pressure and with a high pressure cylinder this can be achieved by the use of a sensitive needle valve or a gas flow regulator. With low pressure cylinders of liquefied gas, particularly where the boiling point of the liquid is close to room temperature, constant pressure output can be achieved by maintaining the cylinder at a constant temperature.[3,57] When this temperature is only slightly higher than that of the boiling point of the liquid, particular care must be taken if a capillary flowmeter or restricting orifice is used. The drop in pressure after the constriction causes rapid cooling and possible

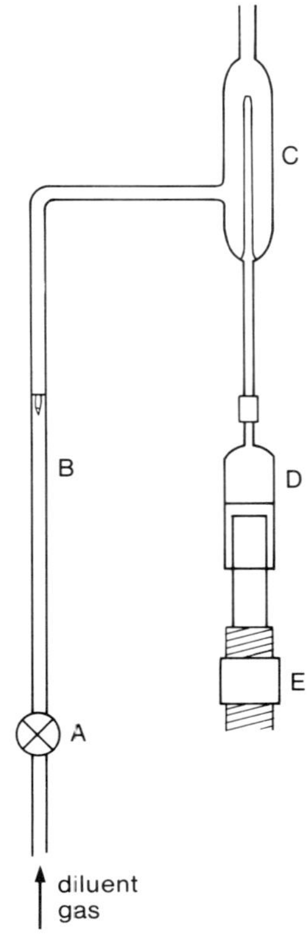

Fig. 7.3 Schematic diagram of apparatus for generating a primary atmosphere by an injection method. A, Flow control valve; B, flowmeter; C, mixing chamber; D, syringe; E, injector ram.

liquefaction, but this may be prevented by heating the restrictor by means of a heating tape or wire-wound heater[3] to maintain it at a temperature above the boiling point. The control of pressure of gases generated by chemical or electrolytic means is usually effected by controlling the reaction rate; this can be difficult and the use of such methods is probably best avoided if at all possible. Gases produced by chemical or electrolytic means may contain impurities derived from the reagents used and secondary reaction products. Means can usually be found to remove these impurities, but this adds to the complexity of the apparatus and introduces further possible sources of error. Ozone is one gas not available commercially in cylinders and may be produced from air or oxygen by a variety of methods of which ultra-violet radiation[48,58] or electrical discharge[59–61] are the most common. To reduce the possibility of the formation of oxides of nitrogen it is usual to use oxygen rather than air.

7.1.2 Injection methods

Injection techniques are convenient for use with liquids and certain gases.[29,62,63] They have also been used successfully for generating atmospheres of volatile solids dissolved in a suitable solvent and injected in solution. The contaminant is added continuously to the diluting gas stream by means of a syringe (Fig. 7.3). Various devices are available for continuous injection at a controlled rate; these are essentially constant speed, electric motors which, through a gear box, drive a ram connected to the plunger of the syringe. These so-called slow injection devices range from expensive models with infinitely variable speeds to relatively cheap models with a limited range of speeds.

To reduce the possibility of diffusion of the diluting gas into the injection jet which might cause possible fluctuations in the concentration of the resultant diluted atmosphere, the diluting gas should flow at a high velocity past the end of the jet. The effect can also be minimised by using a long, capillary, delivery tube with a loosely packed, glass wool plug at the delivery end.[68] Pressure fluctuations within the apparatus, particularly when injecting easily compressible gases, may also result in entrainment of diluting gas within the jet and to avoid this the injection rate should be as high as possible consistent with the required concentration. With liquids, too slow an injection rate may result in little or no atomisation since the liquid may trickle back over the end of the inlet jet and not be forced into the diluting gas stream. To ensure homogeneity of the resultant diluted vapour it is necessary to employ a means of vaporisation of the liquid. This is usually achieved by using an atomiser in which the liquid, in the form of a fine jet, enters the expansion chamber in the same direction as that of the diluting gas stream.[62,69,70] With high-boiling liquids or where high-concentration atmospheres are being generated requiring a fast injection rate, vaporisation may be incomplete and condensation may occur in the chamber causing fluctuations in the concentration of the resultant dilute atmosphere. This may be overcome by maintaining the temperature of the expansion chamber slightly above that of the room by means of a suitable heating jacket. A simple wire-wound heater around the chamber is usually sufficient.[70]

The theoretical concentration of the atmosphere can be calculated from the formula:

$$\text{Concentration} = \frac{10^6 \times v}{Vlt} \text{ p.p.m. by volume at STP for gases}$$

or

$$\frac{10^6 \times v\rho \times 22.4 \times 10^3}{MVlt} \text{ p.p.m. by volume at STP for liquids}$$

where v = total calibrated volume of syringe (cm^3); l = length of syringe over volume v (cm); t = injection rate (cm min^{-1}); V = dilution rate (cm^3 min^{-1}); M = molecular weight of added substance (g); ρ = density of the liquid (g cm^{-3}).

7.1.3 Vapour saturation methods

Vapour saturation provides one of the most generally useful methods for producing dynamic atmospheres of liquids and volatile solids. The diluent gas (sometimes heated) is passed over or through the liquid or solid contained in a suitable vessel (saturator) maintained at a constant temperature. It is often necessary, particularly when dealing with solids and liquids with high boiling points, to use elevated temperatures and some ingenious devices have been developed to achieve and maintain these.[71]

Saturators may take many forms, the commonest being a flask containing the liquid through which the diluting gas is passed via a jetted inlet tube or more often a sintered glass gas-dispersion tube. The efficiency of saturation of the diluting gas will be determined by the bubble size, the residence time of the bubbles in the liquid and the vapour pressure of the liquid. The bubble size is determined by the diameter of the jet or porosity of the dispersion tube and the residence time by the height and viscosity of the liquid column and the flow rate of the gas through the liquid. One major disadvantage of this type of saturator is that the steady loss of liquid causes a change, with time, in the concentration of the generated atmosphere unless some form of self-levelling device is incorporated. However, this loss of liquid can be utilised as a means of calculating the theoretical concentration of the atmosphere from the weight loss over a known period of time. Prolonged passage of air through certain substances may cause oxidation or other chemical reactions and it is usual therefore to use an inert gas, such as nitrogen, rather than air to produce the primary atmosphere.

For certain volatile substances the diluting gas may be blown over the surface of the liquid rather than through it. By this method certain practical problems such as the generation of high back pressures through sintered frits or long columns of liquid may be partly eliminated but incomplete saturation of the diluting gas stream may occur especially with substances of low vapour pressure. This problem may also occur when the gas is bubbled through some liquids. To ensure complete saturation, the effluent, dilute gas stream is cooled rapidly to ensure complete saturation at the new, low temperature; any excess vapour condenses in the cooling vessel.[21,72–75]

Another convenient form of saturator, particularly for solids, consists of a long tube in which is packed a mixture of the substance with some inert material such as kieselguhr or silica;[75–77] this presents a large surface area of the solid to the diluent gas. The liquid or solid may also be dispersed or supported on a paper or glass fibre filter through which the diluent gas is passed.[21,65]

7.1.4 Diffusion methods

Diffusion tubes provide a convenient means of producing atmospheres of low concentration and are particularly suitable for solvent vapours.[78–81] The principle of operation utilises the property of gases and vapours to diffuse at a uniform rate through a tube provided that the temperature and concentration gradient are constant and the tube geometry is unchanged. The diluent gas, usually heated to the same temperature as the diffusion tube, is passed at a constant rate across the mouth of the tube. The resultant atmosphere concentration is controlled by varying the temperature of the diffusion tube or the flow rate of diluting gas. Diffusion tubes with a wide variety of designs have been used[80,82–87] but essentially they consist of a reservoir, containing the substance of interest, connected by means of a capillary tube to a mixing chamber through which the diluting gas is passed (Fig. 7.4). It has been shown that for the best results the ratio of the length of the tube to its cross sectional area should be less than 0.3.[82] Tubes should not be greater than about

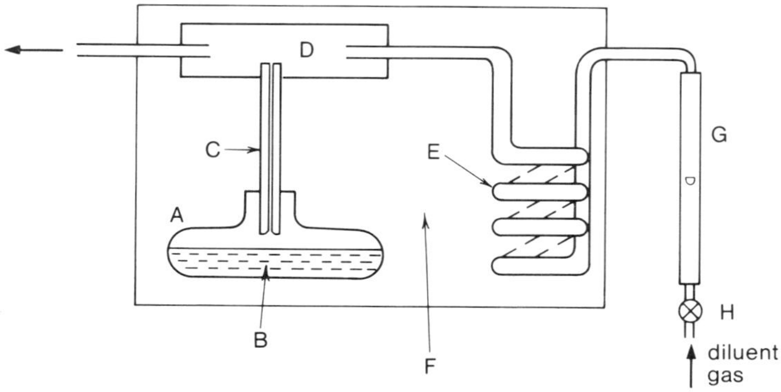

Fig. 7.4 Schematic diagram of apparatus for generating a primary atmosphere by a diffusion method. A, Reservoir; B, contaminant; C, capillary tube; D, mixing chamber; E, metal coil; F, thermostatically controlled bath; G, flowmeter; H, flow control valve.

20 mm in diameter, otherwise turbulence may result which will decrease the effective path length.[82,88]

Diffusion methods provide one of the least complex ways of generating atmospheres of low concentration of the order of 100 p.p.m. or less,[83] although higher concentrations can be achieved. To ensure a uniform diffusion rate, high flow rates for the diluent gas should be avoided. Upper limits of 1 $dm^3\,min^{-1}$ have been quoted,[87,88] but with careful choice of tube dimensions, rates up to 10 $dm^3\,min^{-1}$ may be used, although this is not normally recommended.

The concentration of a dynamic diffusion system can be expressed by

$$\frac{10^6 \cdot r}{V} \text{ p.p.m. by volume at STP}$$

where r = rate of diffusion ($cm^3\,min^{-1}$); V = flow rate of diluent gas ($cm^3\,min^{-1}$).

The diffusion rate (r) can be calculated from the equation:

$$\frac{DA}{L} \ln \frac{P}{P - p_v}$$

where D = diffusion coefficient of the liquid ($cm^2\,s^{-1}$); A = cross sectional area of the diffusion tube (cm^2); L = length of the diffusion tube (cm); V = flow rate of the diluent gas ($cm^3\,s^{-1}$); P = pressure in diffusion cell (atm); p_v = vapour pressure of diffusing vapour (atm).

Tables of diffusion coefficients have been compiled,[89,90] but variations are found and these, together with other experimental factors, often result in wide differences between calculated and observed diffusion rates.[82,87,88,90]

7.1.5 Permeation methods

A simple method for producing low concentration gas mixtures by permeation was suggested by O'Keefe and Ortman.[91] A liquid or pressurised gas is contained in a tube of plastics material sealed at both ends with glass or steel balls. The gas or vapour permeates the walls of the tube at a rate dependent upon the material of the tube, its length, thickness and the temperature at which it is maintained. Once equilibrium has been attained at a particular temperature, the permeation rate will remain constant throughout the life of the tube. The characteristics and performance of permeation tubes have been described in some detail by Lucero.[92] Accurate temperature control is essential since the permeation rate is very temperature dependent[91,93] the rate for sulphur dioxide through FEP Teflon may, for example, double for every 10°C rise and for butane for a similar rise the rate may increase threefold.[91] Permeation tubes provide a convenient means of producing gas mixtures in the parts per million range for the calibration of air pollution analysers. The tubes are suspended in a metered stream of diluent gas and with proper temperature control, high precision and an accuracy of $\pm 2\%$ or better is possible for the resultant mixture.[91,94]

Despite the relative simplicity of the method there are some disadvantages. The number of substances that will permeate fast enough to be useful is limited and this is particularly true of liquids. Some substances react slowly with or in some way affect the permeability of the plastics material used in the construction of the tube; carbon dioxide and nitrogen dioxide have been reported to behave in this way.[95,96] The need for accurate temperature control may necessitate the use of expensive control systems.

The recommended method of storing permeation tubes, to prolong their useful life, is in a refrigerator. However with some plastics materials it has been observed that when they are reheated to the experimental temperature some time is required before the expected permeation rate is reached and this has also been found to have changed, often appreciably.[97] This can be due to absorbed moisture[96,98] and the effect can be reduced by ensuring a similar relative humidity during storage, to that of the experimental conditions.[96] The continuous passage of diluent gas, even at a very low rate, during storage also helps the tube to maintain a constant permeation rate.[98]

Permeation tubes for a number of compounds are now obtainable commercially; these tubes are usually supplied with a certified permeation rate. The manufacture of tubes in the laboratory is possible, but some skill is required and often the expense is not justified.

The tube is probably the most convenient form of permeation device but several other forms have been suggested[99–102] although their performance and characteristics have not been subjected to the same detailed study.

The problem of low permeation rates with some substances has been overcome by increasing the surface area available for permeation by sealing liquids in polyethylene bags[103] rather than plastics tubes. These bags may be made and filled easily and conveniently in the laboratory and the range of concentrations that can be covered is very much wider. Bags have similar disadvantages to those of tubes and also on account of their larger size, storage may be a problem.

The purity of the substances used to fill permeation devices must be high, otherwise erroneous permeation rates may be obtained since the observed rate will depend on the rates of permeation of the individual substances present and the proportion of each in the mixture. The presence of impurities may be suspected if the permeation rate varies with time until it finally reaches a constant value. Permeation devices can therefore provide an effective means of purification.[97]

Permeation methods are particularly useful if it is required to prepare a multi-component gas mixture since several tubes or bags may be introduced into the diluting gas stream at the same time. Mixed standards may be required when assessing the effect of the presence of co-contaminants on a test method or measuring instrument.

7.1.6 Miscellaneous methods and devices

Brief mention should also be made of some other methods and devices that have been used for preparing standard gas mixtures dynamically. These include liquid and gas pistons[21,104–106] and pulse diluters[95] of which the Woestoff gas dosing apparatus is an example. These can be very useful as primary atmosphere sources, but are generally less adaptable for general use than the methods described above. Unstable or highly reactive gases may need to be produced on a small scale, *in situ*, by chemical reaction[1,21,68] or pyrolysis.[102,107,108] Some of the lesser known methods of adding controlled amounts of a substance to a diluting gas stream have been reviewed by Hersch.[21]

7.2 Dilution stage

It is possible, by using diffusion or permeation methods for generation of the primary atmosphere, to obtain the required concentration of the gas mixture without further dilution. However, as a general rule, the preparation of gas mixtures in the parts per million and lower levels requires the use of one or more dilution stages in which the primary atmosphere is mixed with diluting gas in measured amounts; if necessary this dilute mixture may be diluted further.

If possible the number of dilutions should not exceed two or the amount of extra apparatus required becomes unwieldy and is likely to introduce problems with back pressure, turbulence, leakage and sorption. However where a high dilution is necessary it is probably better to achieve this by successive dilutions rather than in a single stage. A convenient ratio of concentrated atmosphere to diluent gas is about 1:1000, but if accurate, high-volume flowmeters are available, ratios of 1:10 000 are possible.

The mixing of gases is usually carried out in mixing chambers. These are small vessels in which the diluting air is introduced at right angles to the incoming concentrated atmosphere. Efficiency of mixing is achieved by the high turbulence induced by the diluting air travelling at high velocity past the tip of the jet carrying the incoming gas. The shape of the vessel itself also induces turbulence and aids mixing. Occasionally it may be necessary to include an extra device such as a baffle-plate mixer to ensure complete homogeneity. Various types of mixing and homogenising devices have been described. Fig. 7.2 shows, schematically, a simple single stage dilution system.

7.2.1 Diluent gas

The diluting gas must be free from any compound which may react with that being generated or have an adverse effect on the materials of construction of the apparatus or the test for which the generated atmosphere is required. The majority of generators are required for use in connection with tests for atmospheric contaminants and therefore the diluting gas will in most cases be air derived in the laboratory from a pressurised cylinder or more often a compressor. In the latter case in particular a number of undesirable impurities may be introduced during compression and storage. These may include oil, carbon dioxide, nitrogen oxides, aldehydes, hydrocarbons and dust particles. Before use, therefore, the air supply must be cleaned by passage through suitable filters. The choice of filter will be determined by the impurities to be removed. Depending upon the likely interference effects, complete purification may be unnecessary and sorbents for specific substances may be adequate. A useful, general purpose method is to pass the air through a glass fibre filter, soda lime, activated charcoal and a second, glass fibre filter or cotton wool plug, in that order. Certain low molecular weight compounds including carbon monoxide may not be removed completely by charcoal and proprietary sorbents. In the case of organic impurities, some form of catalytic pyrolysis may have to be used.[109]

Compressed air supplies are frequently very dry and some means of increasing, controlling and measuring the water content may be desirable to ensure that the generated atmosphere is of the correct relative humidity for the calibration of certain instruments. It is also often necessary to check the effect of humidity on new test methods. In effect any humidifying system will be a secondary, dynamic atmosphere generator using vapour saturation or injection techniques. The former, in which the air is passed over or bubbled through water maintained at a suitable temperature, is probably most commonly used to achieve relative humidities in the range 5–99%, controlled by variation of the air flow rate, temperature or both. For low relative humidities the injection technique using an atomiser[110] (p. 207) is suitable. Measurement of the humidity of the resultant gas stream may be made by use of wet and dry bulb hygrometers, dew-point apparatus, electronic meters and if necessary other specialised hygrometric methods.[111–113]

7.3 Sampling stage

In order that the generated atmosphere may be used for its intended purpose, facility for sampling or attachment of the instrument to be calibrated must be provided. Usually this is a fairly simple matter, but certain precautions are necessary to ensure that the correct sample is taken and that the balance of the generator is not upset. Wherever possible sampling is carried out at atmospheric pressure and a common procedure is to pass the atmosphere into an open bell from which the sample is taken via a suitable probe. To reduce the possibility of local dilution by ambient air, samples are taken from well inside the bell whilst ensuring that the flow rate of the generated atmosphere is well in excess of the sampling rate. When calibrating instruments it may be necessary to provide special sampling facilities. It may, for example, be necessary to place the instrument in a suitably sized chamber through which the generated atmosphere is passed continuously.

Generally the gases and vapours generated will be toxic and adequate precautions must be taken at all times to ensure that there is no health hazard to operators or others. Large volumes of gas mixtures prepared by dynamic methods will often be wasted and vented to the atmosphere; generators should therefore be constructed and used in fume cupboards or under fume hoods. It is advisable particularly when dealing with extremely toxic or carcinogenic substances, that the waste is trapped in a suitable sorbent and disposed of as toxic waste. Inclusion of a trap of this nature can cause problems since, in a completely enclosed system, changes in pressure during sampling may seriously upset flow rates within the generator and possibly lead to variations in the concentration of the atmosphere. This may be overcome by surrounding outlets from the generator with a suitable funnelling tube open to the atmosphere and connected via the sorbent trap to a pump operating at a flow rate very slightly in excess of that of the issuing gas mixture (Fig. 7.2); the effect on the pressure of the system is negligible.

8 Standardisation of the prepared atmosphere

The concentration of a prepared atmosphere can be predicted with reasonable accuracy by calculation. Even with the greatest care, however, there are a number of possible sources of error that can be introduced during the construction of the generator which, particularly at very low concentrations, may have serious effects on the concentration actually obtained. The main sources of error are leaks in the system, losses due to sorption on surfaces and incorrect flowmeter readings caused by the introduction of back pressures. For work of the highest accuracy therefore it is necessary to standardise the atmosphere by means of a suitable chemical or physicochemical method. The method chosen should if possible be quite different from that of the test method being checked or that used in the instrument being calibrated and where possible at least two different methods should be used.

Ideally, standardisation should be carried out at the same time as the test or calibration, but in practice this is rarely necessary particularly with dynamic generators that have been shown to have reached a stable condition. However, fairly frequent checking during use is advisable. Once the precise concentration has been established, much time and effort can be saved by continuously monitoring the generated atmosphere. Devices suitable for this purpose include ionisation, galvanic and photometric detectors provided that they are sensitive to the substance being monitored. Also they must be insensitive to or little affected by air and water vapour, unless some suitable means of 'backing off' is available, and must have a response sufficiently high to make accurate estimation possible.

Flame ionisation detectors are particularly useful for monitoring atmospheres of organic compounds particularly when air is the diluent gas because they are not seriously affected by the presence of water and oxygen as are electron capture detectors. Galvanic detectors may be used for a wide range of compounds;

Table 7.1 The preparation of standard mixtures of gases and vapours by dynamic methods

Compound or class of compound	*Method of preparation** *(pp. 204–211)*	*Reference*
ORGANIC		
Aldehydes	C	117
	E	102
	F	108
Amines		
aliphatic	A	118
aromatic	C	75, 119
	D	79, 119
	E	120
Hydrocarbons		
aliphatic	D	121
aromatic	C	122, 123
	D	79
Isocyanates		
aromatic	B	67
	D	124, 125
Ketones	B	126, 127
	E	103, 127–129
Organohalogen compounds		
aliphatic	C	33, 74, 130, 131
	E	132
Organomercury compounds	B	65
	C	65, 133
	E	133, 134
Phosgene	A	68, 135
	C	136
	E	137, 138
Vinyl chloride	A	139
	E	140–142
INORGANIC		
Ammonia	A	68, 118
	E	143, 144
Carbon disulphide	B	145
	E	146
Carbon monoxide	A	68, 118, 147
	B	148
Halogens	A	68, 118
	C	68
	E	149, 150
	F	151
Hydrogen cyanide	C	68
	D	152
	E	137
Hydrogen fluoride	A	3, 135
	E	153
	F	21
Hydrogen sulphide	D	121
	E	144, 146, 154

Table 7.1 (*continued*)

Compound or class of compound	*Method of preparation** *(pp. 204–211)*	*Reference*
Mercury vapour	C	72, 73, 76, 134, 155–157
	E	156
Nitrogen dioxide	A	68, 118, 158–162
	B	2
	C	60
	D	121
	E	163
	F	164, 165
Sulphur dioxide	A	57, 118, 166
	D	121
	E	94, 143, 146, 167
	F	168
Ozone	A	58–60
Water (Humidity)	B	1, 110
	C	1, 118, 169, 170
	D	171
	E	143

* A: direct dilution; B: injection; C: vapour saturation; D: diffusion; E: permeation; F: miscellaneous.

conductivity meters have, for example, provided a convenient means of monitoring acid gases.

Spectrophotometers are generally more specific in the compounds they can monitor but nevertheless may be extremely useful. Flame photometric detectors are suitable for monitoring compounds containing sulphur, phosphorus and halogens with a high degree of specificity. Atomic absorption is generally more suited to metal fumes rather than gases and vapours, but it is particularly useful for monitoring mercury vapour. Infra-red spectrophotometers also make very useful monitoring devices. Ultra-violet and visible range spectrophotometers can be used as very sensitive monitors. They may, however, be less convenient because of the need to absorb the prepared atmosphere in a suitable solvent to which reagents may have to be added to produce coloured solutions.

Many of the instruments currently used for the detection and measurement of hazardous substances in the atmosphere and described in Chapter 4 may be adapted for use as continuous monitors. However, as with methods used for standardisation, when the generator is used for instrument calibration it is advisable to use a monitor with an operating principle different from that of the instrument being tested.

9 Recommended methods of preparing gas mixtures

It has been the purpose of this chapter to describe the general principles of preparing gas mixtures rather than to recommend a particular method for individual substances. The chosen method will depend on the properties of the substance (see p. 204), the purpose for which the mixture is intended and often on the resources available. In many cases only one method will be suitable, but where a choice exists unnecessary complications should be avoided; the simplest and most straightforward is often the best. Some workers have had to fabricate special and sometimes very ingenious devices to generate the primary atmosphere, but generally this has been necessary because no other suitable method was available.

The wide availability of pure gases in pressurised cylinders has eliminated the need, in a large number of cases, to use chemical methods to generate the starting product. The direct dilution method of preparation can therefore be applied more generally and easily than was the case a few years ago. Similarly a better understanding of the problems involved in preparing stable, dilute gas mixtures in cylinders or aerosol cans has meant that commercially available products may be used with greater confidence as to their concentration than in the past. There should be even greater improvement and reliability as standards for these dilute mixtures, which are particularly suited for the calibration of instruments, become internationally accepted.

Methods for the preparation of mixtures of selected gases and vapours have been reviewed.[1,68,114–116] Table 7.1 lists references to papers describing dynamic methods for the preparation of standard gas mixtures of a number of individual substances or classes of compounds; the list is not exhaustive.

10 References

1. G. O. Nelson. *Controlled Test Atmospheres.* Ann Arbor Science Publishers, Inc., Ann Arbor, Michigan (1971).
2. B.S.S. 4559, British Standards Institution, London (1970).
3. B. S. Marshall and R. Wood. *Analyst,* **93**, 821 (1968).
4. G. Barr. *J. Scient. Instrum.* **11**, 321 (1934).
5. A. Levy. *Ibid.* **41**, 449 (1964).
6. J. J. Czubryt and H. D. Gesser. *J. Gas Chromat.* **6**, 528 (1968).
7. J. J. Hunter. *J. Scient. Instrum.* **42**, 175 (1965).
8. G. J. Frisone. *Chemist-Analyst,* **54**, 56 (1965).
9. A. F. Brown and H. Kronberger. *J. Scient. Instrum.* **24**, 151 (1947).
10. J. E. Lovelock and E. M. Wasilewska. *Ibid.* **26**, 367 (1949).
11. T. Hatch, H. Warren and P. Drinker. *J. Ind. Hyg. Toxicol.* **14**, 301 (1932).
12. R. T. Page. *Ind. Engng. Chem.* **7**, 355 (1935).
13. J. W. Anderson and R. Friedman. *Rev. Scient. Instrum.* **20**, 61 (1949).
14. M. Corn and W. Bell. *Amer. Ind. Hyg. Assoc. J.* **24**, 502 (1963).
15. J. P. Lodge, J. B. Pate, B. E. Ammons and G. A. Swanson. *J. Air Poll. Control Assoc.* **16**, 197 (1966).
16. C. Huygen. *Ibid.* **20**, 675 (1970).
17. D. L. Brenchley. *Ibid.* **22**, 967 (1972).
18. P. Kotrappa, N. S. Pimpale, P. S. S. Subrahmanyam and P. P. Joshi. *Ann. Occup. Hyg.* **20**, 189 (1977).
19. D. W. Bradley. *Chemist-Analyst,* **55**, 93 (1966).

20. H. D. Axelrod, J. H. Carey, J. E. Bonnelli and J. P. Lodge. *Analyt. Chem.* **41**, 1856 (1969).
21. P. A. Hersch. *J. Air Poll. Control Assoc.* **19**, 164 (1969).
22. H. A. Druett. *Brit. J. Ind. Med.* **12**, 65 (1955).
23. B. E. Saltzman and N. Gilbert. *Amer. Ind. Hyg. Assoc. J.* **20**, 379 (1959).
24. C. B. Avera. *Rev. Scient. Instrum.* **32**, 985 (1961).
25. J. Novak and J. Janak. *J. Scient. Instrum.* **38**, 374 (1961).
26. J. F. Byrd. *J. Air Poll. Control Assoc.* **7**, 58 (1957).
27. D. M. Benforado, W. J. Rotella and D. L. Horton. *Ibid.* **19**, 101 (1969).
28. F. M. Stead and G. J. Taylor. *J. Ind. Hyg. Toxicol.* **29**, 408 (1947).
29. H. N. Cotabish, P. W. McConnaughey and H. C. Messer. *Amer. Ind. Hyg. Assoc. J.* **22**, 392 (1961).
30. A. G. Apol, W. A. Cook and E. F. Lawrence. *Ibid.* **27**, 149 (1966).
31. H. P. Williams and J. D. Winefordner. *J. Gas Chromat.* **4**, 271 (1966).
32. R. P. DeGrazio. *Ibid.* **6**, 468 (1968).
33. E. E. Campbell, M. F. Milligan and H. M. Miller. *Ibid.* **20**, 138 (1959).
34. R. S. Brief and F. W. Church. *Amer. Ind. Hyg. Assoc. J.* **21**, 239 (1960).
35. E. D. Baretta, R. D. Stewart and J. E. Mutchler. *Ibid.* **30**, 537 (1969).
36. G. Leonardos, D. Kendall and N. Barnard. *J. Air Poll. Control Assoc.* **19**, 91 (1969).
37. R. O. McCaldin and E. R. Hendrikson. *Ibid.* **20**, 509 (1959).
38. M. J. Morley and B. D. Tebbens. *Amer. Ind. Hyg. Assoc. Quart.* **14**, 303 (1953).
39. R. S. Brief, R. S. Ajemian and R. G. Confer. *Amer. Ind. Hyg. Assoc. J.* **28**, 21 (1967).
40. K. W. Wilson and H. Buchberg. *Ind. Engng. Chem.* **50**, 1705 (1958).
41. A. N. Setterlind. *Amer. Ind. Hyg. Assoc. Quart.* **14**, 113 (1953).
42. J. Feldstein, J. D. Coons, H. C. Johnson and J. E. Yocum. *Amer. Ind. Hyg. Assoc. J.* **20**, 374 (1959).
43. S. Russell. *Ibid.* **25**, 359 (1964).
44. D. N. Hanson and A. Maimoni. *Analyt. Chem.* **31**, 158 (1959).
45. S. P. Vango. *Chemist-Analyst,* **52**, 53 (1963).
46. P. Urone, J. B. Evans and C. Mo. Noyes. *Analyt. Chem.* **37**, 1104 (1965).
47. J. R. Brown and E. Mastromatteo. *Amer. Ind. Hyg. Assoc. J.* **25**, 560 (1964).
48. W. D. Conner and J. S. Nader. *Ibid.* **25**, 291 (1964).
49. H. C. Wohlers, N. M. Trieff, H. Newstein and W. Stevens. *Atmos. Envir.* **1**, 121 (1967).
50. R. H. Groth and T. B. Doyle. *J. Gas Chromat.* **6**, 138 (1968).
51. W. E. Gill. *Amer. Ind. Hyg. Assoc. J.* **21**, 87 (1960).
52. R. A. Baker and R. C. Doerr. *Int. J. Air Poll.* **2**, 142 (1959).
53. A. P. Altschuller, A. F. Wartburg, I. R. Cohen and S. F. Sleva. *Int. J. Air Wat. Poll.* **6**, 75 (1962).
54. C. A. Clemons and A. P. Altschuller. *J. Air Poll. Control Assoc.* **14**, 407 (1964).
55. F. J. Shuette. *Atmos. Envir.* **1**, 515 (1967).
56. B. S. Smith and J. O. Pierce. *Amer. Ind. Hyg. Assoc. J.* **31**, 343 (1970).
57. N. Birks and T. Flatley. *J. Scient. Instrum.* **2**, 436 (1969).
58. C. H. Thienes, R. G. Skillen, A. Hoyt and E. Bogen. *Amer. Ind. Hyg. Assoc. J.* **26**, 255 (1965).
59. R. H. Hendriks and L. B. Larsen. *Ibid.* **27**, 80 (1966).
60. I. C. Cohen, A. F. Smith and R. Wood. *Analyst,* **93**, 507 (1968).
61. S. Kaye and J. E. Koency. *Rev. Scient. Instrum.* **40**, 505 (1969).
62. D. W. Hill and H. A. Newell. *J. Scient. Instrum.* **42**, 783 (1965).
63. G. O. Nelson and K. S. Griggs. *Rev. Scient. Instrum.* **39**, 927 (1968).
64. W. M. Diggle and J. C. Gage. *Analyst,* **78**, 473 (1953).
65. A. A. Christie, A. J. Dunsdon and B. S. Marshall. *Ibid.* **92**, 185 (1967).
66. D. W. Meddle and R. Wood. *Chem. Ind.* 1635 (1968).
67. D. W. Meddle, D. W. F. Radford and R. Wood. *Analyst,* **94**, 369 (1969).

68. B. E. Saltzman. *Analyt. Chem.* **33**, 1100 (1961).
69. J. C. Gage. *J. Scient. Instrum.* **30**, 25 (1953).
70. B. S. Marshall. *Ibid.* **43**, 119 (1966).
71. M. Krzymien and L. Elias. *J. Phys. (E) Scient. Instrum.* **9**, 584 (1976).
72. G. A. Sergeant, B. E. Dixon and R. G. Lidzey. *Analyst,* **82**, 27 (1957).
73. G. O. Nelson. *Rev. Scient. Instrum.* **41**, 776 (1970).
74. C. P. Roper. *Amer. Ind. Hyg. Assoc. J.* **32**, 847 (1971).
75. *Report of the Government Chemist 1970,* p. 101. HM Stationery Office, London (1971).
76. *Report of the Government Chemist 1974,* p. 99. HM Stationery Office, London (1975).
77. P. A. Pella. *Analyt. Chem.* **48**, 1632 (1976).
78. J. S. Nader. *Amer. Ind. Hyg. Assoc. J.* **19**, 1 (1958).
79. E. A. Campbell and H. M. Ide. *Ibid.* **27**, 323 (1966).
80. A. Dravnieks and B. K. Krotoszynski. *J. Gas Chromat.* **6**, 144 (1968).
81. H. Arito and R. Soda. *Ind. Health (Kawasaki, Jap.),* **6**, 120 (1968).
82. J. M. H. Fortuin. *Analyt. Chim. Acta,* **15**, 521 (1956).
83. J. M. McKelvey and H. E. Hoelscher. *Analyt. Chem.* **29**, 123 (1957).
84. J. E. Lovelock. *Ibid.* **33**, 162 (1961).
85. G. A. Lugg. *Ibid.* **41**, 1911 (1969).
86. A. C. Savitsky and S. Siggia. *Ibid.* **44**, 1712 (1972).
87. A. M. Miguel and D. F. S. Natusch. *Ibid.* **47**, 1705 (1975).
88. A. P. Altschuller and I. R. Cohen. *Ibid.* **32**, 802 (1960).
89. *International Critical Tables,* Vol. 5. McGraw-Hill, New York (1929).
90. G. A. Lugg. *Analyt. Chem.* **40**, 1072 (1968).
91. A. E. O'Keefe and G. C. Ortman. *Ibid.* **38**, 760 (1966).
92. D. P. Lucero. *Ibid.* **42**, 1744 (1971).
93. B. E. Saltzman. *J. Air Poll. Control Assoc.* **18**, 326 (1968).
94. C. E. Rodes, H. F. Palmer, L. A. Elfers and C. H. Norris. *Ibid* **19**, 575 (1969).
95. H. D. Axelrod, J. B. Pate, W. R. Barchet and J. P. Lodge. *Atmos. Envir.* **4**, 208 (1970).
96. B. E. Saltzman, W. R. Burg and G. Ramaswami. *Envir. Sci. Technol.* **32**, 299 (1971).
97. S. A. Stern, T. F. Sinclair, P. J. Gareis, N. P. Vahldieck and P. H. Mohr. *Ind. Engng. Chem.* **57**, 49 (1965).
98. P. P. Scaringelli, A. E. O'Keefe, E. Rosenburg and J. P. Bell. *Analyt. Chem.* **42**, 871 (1970).
99. J. M. Hales, D. A. Snyder and J. L. York. *Amer. Ind. Hyg. Assoc. J.* **32**, 299 (1971).
100. F. Lindqvist and R. W. Lanting. *Atmos. Envir.* **6**, 943 (1972).
101. R. N. Crossett and D. J. Campbell. *Lab. Pract.* **23**, 369 (1974).
102. J. Godin, G. Bouley and Cl. Boudene. *Analyt. Lett.* **A11**, 319 (1978).
103. P. Andrew and R. Wood. *Chem. Ind.* 1836 (1968).
104. H. F. Calcote. *Analyt. Chem.* **22**, 1058 (1950).
105. L. G. Lundsted, A. B. Ash and N. L. Koslin. *Ibid.* **22**, 626 (1950).
106. A. Charnley, G. L. Isles and J. S. Robinson. *J. Scient. Instrum.* **31**, 145 (1954).
107. W. Tsang. *J. Res. Nat. Bur. Stand., Sect. A.* **78**, 157 (1974).
108. W. Tsang and J. A. Walker. *Analyt. Chem.* **49**, 13 (1977).
109. H. L. Kusnetz, B. E. Saltzman and M. E. Lanier. *Amer. Ind. Hyg. Assoc. J.* **21**, 361 (1960).
110. D. W. Hill and N. A. Newell. *Nature, London,* **205**, 593 (1965).
111. J. N. Crawshaw and H. G. Davidson. *J. Scient. Instrum.* **36**, 121 (1959).
112. H. A. Keidel. *Analyt. Chem.* **31**, 2043 (1959).
113. J. A. J. Walker and P. Campion. *Analyst,* **90**, 199 (1965).
114. British Occupational Hygiene Society, Report of Working Group on Standard Atmospheres; private communication.
115. W. J. Becker. *Staub-Reinhalt. Luft,* **37**, 183 (1977).
116. W. J. Becker. *Ibid.* **37**, 278 (1977).

117. W. E. Rinehart. *Amer. Ind. Hyg. Assoc. J.* **28**, 561 (1967).
118. G. O. Nelson, L. E. Swisher, R. D. Taylor and B. E. Bigler. *Ibid.* **36**, 49 (1975).
119. *Report of the Government Chemist 1976,* p. 80. HM Stationery Office, London (1977).
120. *Report of the Government Chemist 1972,* p. 94. HM Stationery Office, London (1973).
121. J. H. Blacker and R. S. Brief. *Amer. Ind. Hyg. Assoc. J.* **32**, 668 (1971).
122. A. J. Blake and B. A. Rose. *Analyst,* **85**, 442 (1960).
123. R. M. Ash and J. R. Lynch. *Amer. Ind. Hyg. Assoc. J.* **32**, 410 (1971).
124. A. Neilson and K. S. Booth. *Ibid.* **36**, 169 (1975).
125. *Report of the Government Chemist 1978,* p. 98. HM Stationery Office, London (1979).
126. A. F. Smith and R. Wood. *Analyst,* **95**, 683 (1970).
127. A. F. Smith and R. Wood. *Ibid.* **97**, 363 (1972).
128. P. Andrew and R. Wood. *Ibid.* **95**, 691 (1970).
129. P. Andrew, A. F. Smith and R. Wood. *Ibid.* **96**, 528 (1971).
130. G. J. Stopps and M. McLaughlin. *Amer. Ind. Hyg. Assoc. J.* **28**, 43 (1967).
131. R. M. Ash and J. R. Lynch. *Ibid.* **32**, 552 (1971).
132. H. B. Singh, L. Salas, D. Lillian, R. R. Arnts and A. Appleby. *Envir. Sci. Technol.* **11**, 511 (1977).
133. A. L. Lynch, R. F. Stalzer and D. T. Lefferts. *Amer. Ind. Hyg. Assoc. J.* **29**, 79 (1968).
134. P. E. Trujillo and E. E. Campbell. *Analyt. Chem.* **47**, 1629 (1975).
135. P. A. Pella, E. E. Hughes and J. K. Taylor. *Amer. Ind. Hyg. Assoc. J.* **36**, 755 (1975).
136. W. E. Rinehart and T. Hatch. *Ibid.* **25**, 545 (1964).
137. A. W. Barendsz and G. van der Linden-Tak. *Z. analyt. Chem.* **274**, 207 (1975).
138. G. G. Esposito, D. Lillian, G. E. Podolak and R. M. Tuggle. *Analyt. Chem.* **49**, 1774 (1977).
139. B. Miller, P. O. Kane, D. B. Robinson and P. J. Whittingham. *Analyst,* **103**, 1165 (1978).
140. J. E. Cuddeback, W. R. Burg and S. R. Birch. *Envir. Sci. Technol.* **9**, 1168 (1975).
141. W. R. Burg, S. R. Birch, J. E. Cuddeback and B. E. Saltzman. *Ibid.* **10**, 1233 (1976).
142. L. H. Nelms, K. D. Reiszner and P. W. West. *Analyt. Chem.* **49**, 994 (1977).
143. K. A. Rehme. *Ibid.* **47**, 2474 (1975).
144. A. Teckentrup and D. Klockow. *Ibid.* **50**, 1728 (1978).
145. E. C. Hunt, W. A. McNally and A. F. Smith. *Analyst,* **98**, 585 (1973).
146. R. R. Stevens, A. E. O'Keefe and G. C. Ortman. *Envir. Sci. Technol.* **3**, 652 (1969).
147. A. S. Morgenstern, R. M. Ash and J. R. Lynch. *Amer. Ind. Hyg. Assoc. J.* **31**, 630 (1970).
148. L. Silverman and G. R. Gardner. *Ibid.* **26**, 97 (1965).
149. B. A. Johnson and C. P. Roper. *Ibid.* **33**, 533 (1972).
150. J. Gabbay, M. Davidson and A. E. Donagi. *Analyst,* **101**, 128 (1976).
151. J. D. Ludwig. *Analyt. Chem.* **41**, 1907 (1969).
152. B. E. Dixon, G. C. Hands and A. F. F. Bartlett. *Analyst,* **83**, 199 (1958).
153. L. A. Elfers and C. E. Decker. *Analyt. Chem.* **40**, 1659 (1968).
154. B. A. Johnson. *Amer. Ind. Hyg. Assoc. J.* **33**, 811 (1972).
155. E. P. Scheide and J. K. Taylor. *Envir. Sci. Technol.* **8**, 1097 (1974).
156. G. L. Corte, G. Dowd and L. Dubois. *Amer. Ind. Hyg. Assoc. J.* **36**, 873 (1975).
157. C. S. McCammon and J. W. Woodfin. *Ibid.* **38**, 378 (1977).
158. B. E. Saltzman. *Analyt. Chem.* **26**, 1949 (1954).
159. B. E. Saltzman and A. F. Wartburg. *Ibid.* **37**, 1261 (1965).
160. F. Nietruch and K. E. Prescher. *Z. analyt. Chem.* **244**, 294 (1969).
161. R. Di. Martini. *Analyt. Chem.* **42**, 1102 (1970).
162. J. E. Higuchi, A. Hsu and R. D. MacPhee. *J. Air. Pollut. Cont. Assoc.* **26**, 136 (1976).
163. F. P. Scaringelli, E. Rosenberg and K. A. Rehme. *Envir. Sci. Technol.* **4**, 924 (1970).

164. J. T. Shaw. *Atmos. Envir.* **1**, 81 (1967).
165. A. A. Christie, R. G. Lidzey and D. W. F. Radford. *Analyst,* **95**, 519 (1970).
166. W. G. Cummings and M. W. Redfearn. *Chem. Ind.* 809 (1957).
167. R. M. Ash and J. R. Lynch. *Amer. Ind. Hyg. Assoc. J.* **32**, 490 (1971).
168. D. Cornell and W. Tsang. *Analyt. Chem.* **46**, 333 (1974).
169. F. E. Harris and L. K. Nash. *Ibid.* **23**, 736 (1951).
170. V. L. Gelizunas. *Ibid.* **41**, 1400 (1969).
171. A. Goldup and M. T. Westaway. *Ibid.* **38**, 1657 (1966).

Index